What others are saying about this book:

"An informative and well presented How-to-Book for those considering starting this fascinating hobby with mason bees."

—**Dr. M. Winston, Simon Fraser University, BC Canada**

"The backyard gardener will find this book filled with detailed information on how to manage and have fun with mason bees".

—**Dr. J. Cane, Logan University, Utah**

"This book is an invaluable tool for anybody wishing to know more about mason bees. I would recommend this book to anybody who is interested in achieving better pollination in their backyard or to somebody who is interested in knowing more about mason bees."

—**S. Devlin, M.P.M, biologist**

"One can easily follow the necessary directions on how to develop mason bee population. Included is information on some mason bee relatives: honey bees, bumble bees and leaf cutter bees."

—**D. McCutcheon, beekeeper**

"The wonderful illustrations make it possible for even the urban gardener to enjoy a harvest of their favourite fruit from their yard."

—**E. Benndorf, 2001 Nature garden winner**

"Nifty, easy to read manual.

—**M. Roboz, M.P.M, gardener**

Pollination
with
Mason Bees

A Gardener's Guide to
Managing Mason Bees
for Fruit Production

MARGRIET DOGTEROM

Second Edition
Fully Revised

Beediverse Publishing®
Coquitlam, British Columbia

Pollination with Mason Bees
A Gardener's Guide to Managing
Mason Bees for Fruit Production
by Dr. Margriet Dogterom

Published by:
Beediverse Publishing of CPC Ltd.
Coquitlam, BC, Canada
info@beediverse.com
http://beediverse.com
Fax: 1-604-936-3927
Tel: 1-800-794-2144

Copyright 2002, 2005, 2009
Cover design © 2002, 2009 Wen Lin & Margriet Dogterom
Cover photos of bees © 2009 Cliff Forest
Line drawings © 2002 Andrea Lonon
Drawings of Beetles © 2009 Margriet Dogterom

ISBN 0-9689357-0-2
First Edition 2002
Second Printing 2005
Second Edition 2009, completely revised
Second Printing 2013
Printed in Canada   Canadian   Softcover

Library and Archives Canada Cataloguing in Publication

Dogterom, Margriet, 1946-
Pollination with mason bees: a gardener's guide to
managing mason bees for fruit production / Margriet
Dogterom. -- 2nd edition. completely rev.

Includes biliographical references and index.
ISBN 978-0-9689357-3-6

1. Orchard Mason bee. 2. Bee culture. I. Title.

SF539.8.O73D63 2009   638'.12   C2009-905572-4

Table of Contents

About the Author

Dr. Margriet Dogterom is the founder of Beediverse® Products of CPC Ltd.. She is an internationally recognized scientist, beekeeper, teacher and author of educational materials about mason bees.

She has designed systems that work to bring the native North American mason bee and its pollinating effectiveness to countless numbers of home gardeners and orchardists. Her innovative Stacked nesting Tray System (STS) makes mason bee keeping easier than ever before and allow mason bees to thrive and stay healthy from one year to the next.

Dr. Dogterom was born in the Netherlands in 1946. She emigrated with her family to Australia in 1957 and to Canada in 1969. She earned her B.Sc., her M.Sc. in honey bee biology and her Ph.D. in pollination, at Simon Fraser University, B.C., Canada.

For the past 20 years, Dr. Dogterom has devoted her time to field research with mason bees, honey bees, bumble bees and leafcutter bees. Her Ph.D. thesis involved the management of these four bee species for crop pollination. She continues her research on mason bee management systems and the control of mason bee pests.

For a detailed list of Beediverse® mason bee housing and educational materials please visit www.beediverse.com.

Preface

 The idea of writing this book came from enthusiastic gardeners and their question "Now what do I do with my mason bees?" This is the first book on mason bees written for the layperson that covers how to manage mason bees, and how to build nests for them.

The decline of honey bee populations, subsequent poor pollination and reduced fruit yields, has lead to a demand for mason bees. Well managed mason bees can pollinate back yard gardens or commercial operations. This book uses the old name "mason bee" for the early spring pollinator *Osmia lignaria* Cresson. Mason bee aptly describes how the bees gather mud to plug their nests. In North America, is it also known as "orchard mason bee" and "blue orchard bee". Despite the term, mason bees can pollinate any crop, not just orchards.

This book is organized as a hands-on resource. All the chapters have been revised for the second edition. It contains practical information on caring for, sustaining, and increasing the number of mason bees in your garden.

- Chapter 1 provides information on pollination, the importance of pollination and how to distinguish bees from wasps and flies.
- Chapters 2 gives information on the mason bee life cycle and where the bees nest naturally without human intervention.

- Chapter 3 contains information on how to get started with a nest.
- Chapter 4 describes various nesting materials including wood, and the type of nest that can be made out of these materials.
- Chapter 5 is an extension of chapter 3 with more detail on where to place nests, on orientation cues, and how to protect nests.

Fall and winter management has been split into two chapters.

- Chapter 6 gives a detailed description of harvesting and cleaning mason bee cocoons. These methods vary depending on the type of nests used.
- Chapter 7 describes the care and storage of cleaned cocoons.
- Chapter 8 is a new chapter which includes details about pests, parasites and predators.
- Chapter 9 gives background information on mason bees, other solitary bees and social bees.
- Chapter 10 is a new chapter describing some novel ideas.
- Chapter 11 is a new chapter on projects.
- Chapter 12 is a chart for bee management throughout the year.

The purpose of this book is to guide the reader into the world of bees and in particular, mason bees.

Acknowledgments

My sincere thanks go to all my friends, associates and workshop participants who encouraged me to write this book on how to look after mason bees. The chapters of this book grew out of the many questions I answered on the web, over the phone and at my mason bee workshops. My thanks to you all.

In particular I would like to thank the following contributors to the first edition: Victoria Bennett, Jordi Bosch, Mike Burgett, Jim Cane, Irm Dogterom, Tim Frizzell, Ernie Fuhr, Magna Glosli, Gordon Kern, Jenny Lakeman, Wen Lin, Andrea Lonon, Melissa Martiny, Patricia Rathbun, Mona Reaume, Bob Smith, Rosetta Smith, Sally Sprague, Bill Stephen, Bertha Verstegen, John Vigna, David Ward and Rex Welland.

Also, a thank you to Henri Fabre who captivated my attention and peaked my curiosity with his delightful writing on solitary bees.

For the 2nd edition, fully revised, I would like to thank the following: Ella Benndorf, Elizabeth Croft, Steve Dupey, Cliff Forest, Tim Frizzell, Terry Griswold, Wen Lin, John Macdonald, Alex Mann, Chris O'Toole, Randy Person, Dick Scarth, Jilian Scarth, Karen Strickler, Rex Welland, Paul Van Westendorp, and Karen Young.

It is with thanks that I include Jim Gaskin of Gaskin Farms, who from the very start of my research on blueberry pollination, said that we can only progress with agriculture through research. He continues to support research and development in pollination and mason bees.

All have contributed, but my ideas came from the experience of keeping bees and in particular mason bees.

Bee photos on cover.

Front cover photo by Margriet Dogterom:
Female mason bee on apple flower

Back cover bee photo by Cliff Forest:
upper- female mason bee flying to flowers

Back cover bee photo by Cliff Forest: lower-
Two male mason bees with their distinguishing white whiskers.

Chapter 1
Pollination and Bees

Pollination of flowers is mostly carried out by bees, although other foragers such as birds also pollinate flowers. In this chapter, bee characteristics and the pollination process are explained.

If you are a gardener you may have noticed a decline in your fruit production. In North America, during the late 1990s and until recently, in many city and suburban gardens, fewer bees means inadequate pollination and has been the cause for smaller fruit. This is because we have depended on the wild honey bee for pollination and now their colonies have been decimated by two parasitic mites (Varroa and tracheal mites) as well as other diseases and pests. It is clear that without wild honey bee colonies there are insufficient bees to pollinate the many blossoms on our fruit trees.

Most honey bee colonies are managed by beekeepers outside the city. In addition, many municipalities prohibit beekeeping within municipal boundaries.

Fortunately, native mason bees can help restore the former productivity of your fruit trees, blueberries and raspberries within the suburbs. Simply provide a nest for mason bees and they will reproduce and pollinate your fruit trees. This book describes how to house and manage mason bees.

From plentiful fruit to few fruit.

1.1 WHAT IS POLLINATION?

Simply, pollination is the transfer of pollen from one flower to another. When more pollen is delivered to a flower, fruit becomes larger. More specifically pollen is transferred from the male part of a flower, the anther, to the receptive stigma on the female part of a flower (pistil). With bees, pollination occurs when they visit flowers to gather pollen and nectar. Once the pollen is delivered to the flower, pollen grains grow pollen tubes down the stalk of the pistil. The genetic material from pollen is transported via the pollen tube to the female ovary where fertilization takes place.

Pollination is the transfer of pollen from one flower to another.

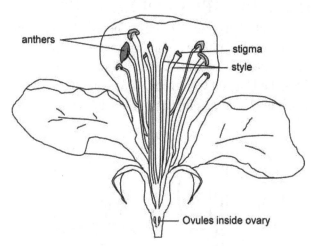

Longitudinal section of an apple flower
(Redrawn from McGregor 1976).

A very close relationship exists between flowers and bees. Bees obtain their food from flowers, and many flowers depend on bees and other animals for pollination. A few flowers are so specialized they can only be pollinated by one species of bee, but generally flowers are pollinated by many bee species. This is because flowers and bees have evolved together over millions of years. The flower produces pollen and nectar that attracts the bee, thus ensuring pollination. The bee receives food as a reward, ensuring that the bee can reproduce.

Pollen grains are needed for fertilization, and the amount of pollen helps determine how big the fruit will become. After fertilization, the fertilized ovule develops into a seed. Thus, seeds in fruit are the result of fertilization after pollination. Fruit develops in response to the hormonal stimulation of fertilization. In other words, more bees = more pollen delivered = more fertilization = more seeds = more fruit and larger fruit.

1.2 HOW DO BEES POLLINATE?

Bees are insects that collect pollen on their pollen-carrying hairs. These branched and plumose hairs are present over many of their body parts and allow bees to collect pollen from flowers. Without these branched hairs it would be more difficult to collect pollen. The remaining hairs are simple hairs (not branched).

Simple hairs   *Branched hairs.*

Back at the nest, pollen is used to feed and nourish young developing bees. Pollen is the 'meat' that provides proteins and fats to build and maintain body structure of bees.

Bees have a variety of specialized structures to carry pollen. Pollen is collected amongst the hairs of a bee and groomed into specialized structures such as pollen baskets or corbiculae. Honey bees and bumble bees pack pollen into a pollen basket located on their hind legs. Mason bees pack pollen amongst stiff hairs called scopa located underneath their abdomen. To read more about honey bees and bumble bees, see the chapter on the relatives of mason bees (Chapter 9).

Nectar is produced at the base of a flower. Bees search for nectar by probing flowers and quickly learn the easiest way to obtain nectar from the often hidden nectaries. Nectar contains sugar, which is the bee's energy source. While probing flowers for nectar, bees usually come in contact with pollen.

Bees are distinguished by three body segments, six legs and four scale-free wings. Their body consists of a head, containing the eyes, mouth-parts and antennae. Wings and legs are attached to the middle part of the body or thorax. The third body segment consists of the abdomen. The female bee has an ovipositor, for laying eggs, which also functions as a stinger. The male bee has male genitalia instead of an ovipositor. Thus no male bee can sting.

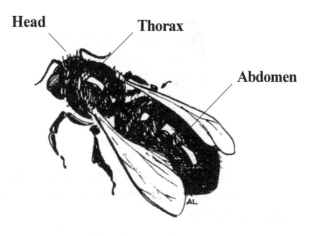

Female mason bee

In general, mason bees appear black. In sunlight, mason bees appear black with an iridescent blue sheen. There are 3 features that distinguish male bees: size, white facial tufts and long antennae. White tufts of hair are visible on newly emerged male bees' faces and can be used to distinguish males from females (see photos on back cover of this book). Males are usually smaller than females and emerge first in spring. The male feature that easily separate it from females is their antennae length. Male mason bees have long antenna, longer than the length of their head. Female mason bees have shorter antenna that are about the same length as their head.

1.3 IS IT A BEE, A WASP OR A FLY?

Bees and wasps are closely related insects that evolved from a common ancestor. The major difference between bees and wasps is that bees, with their specialized branched and plumose hairs, obtain their protein from pollen and are vegetarian; wasps are carnivorous and obtain their protein from meat and other insects. Wasps appear relatively hairless.

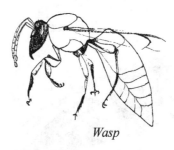

Wasp

Similarly to bees, wasps forage for sugar from flowers and in addition forage for sugar from pop bottles, decaying fruit or any other sugary substance.

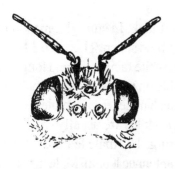

Bees have rod-like, segmented antennae

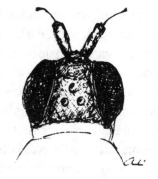

Flies have one thin bristle attached to each antenna club

Flies often mimic bees in colour and shape, but they can be distinguished from bees and wasps by having a close look at their antennae. Bees have a pair of antennae made of rodlike segments. Flies often mimic the striped appearance of bees, but the absence of rodlike segmented antennae is a giveaway that you are looking at a fly and not a bee, no matter what its colour. Bluebottle flies look a lot like mason bees, but their habit of being on refuse and not on open flowers is a good indicator you are looking at a fly and not a bee.

YOUR NOTES:

_____

_____

_____

_____

_____

_____

_____

_____

_____

_____

_____

Chapter 2
Mason Bees

Mason bees are solitary bees grouped within the Genus *Osmia*. Solitary means that each female nests without the cooperation of other female bees. The female chooses a nest, provisions the nest and produces the young by herself. This Chapter covers their lifecycle, nest and food requirements.

Mason bee on an apple blossom

Mason bees are usually smaller than a honey bee. They are black or metallic blue-green, sometimes brilliant. While they are common in western North America, they are rare in deserts, moderately common in eastern North America and uncommon in Mexico. There are about 135 species of mason bees (Chapter 9).

The most common mason bee species is *Osmia lignaria* Cresson, also known as the mason bee, the blue orchard bee, or orchard mason bee. It is active in the early spring.

Since the 1970s, research has determined that the North American mason bee can be managed for pollinating crops. This is possible because mason bee populations can be increased in number and their parasites and predators controlled. Mason bees successfully pollinate tree fruits such as apple, cherry and almond, and berry bushes such as blueberry and raspberry.

Mason bees are fun to watch. The enjoyable writings by the naturalist, J. Henry Fabre (1916) describes the delights in watching bees. When teaching, Henry Fabre took his students outdoors to teach them trigonometry. It was the students' favourite course because it meant a day roaming the countryside, albeit in straight lines. He would send his students out to a point, pacing as they travelled, but his students kept on bending down and sticking items in their mouths. Of course this meant that the trigonometry lesson was forgotten. Fabre investigated these antics and found to his pleasure that his students were collecting nectar and pollen from the 'mason bees of the stones'. These mason bees built their mud nests on indentations and small cavities in the surface of rocks. Within these mud nests his students had found small lumps of pollen and nectar.

*Two children have a close up view of the
friendly mason bees at their nests.*

2.1 LIFE STAGES

Adult mason bees overwinter inside their cocoons until early spring, emerge from their cocoon, mate, produce their offspring for about a month and then die.

1. Spring

Mason bees are active during a distinct time period from February to May depending on latitude and elevation. In spring, adult bees chew their way out of their cocoons. Males are in the outer chambers of the nesting tunnels and emerge first. They can easily be recognized with a tuft of white hairs on the front of their face and their long antennae. After they emerge, they remain around the tunnel entrances, flying very little while they wait for the females to emerge.

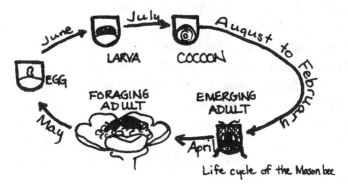

Life cycle of the Mason bee

Female mason bee emerging from her nesting tunnel

After several days of warm weather, females begin to emerge from the deeper chambers of the nesting tunnel. The female can be distinguished from the male by the absence of a white tuft of hair on the front of her face. She also has shorter antennae, hairs under her abdomen and, is larger than the male. Once they emerge, they become active when daytime temperatures reach 14°C (57°F).

Several after days of emerging, the male is sexually mature and immediately mates with the newly emerged virgin female. After mating, the female bee searches for a suitable nesting site. Once she finds a new nest cavity, she begins foraging for pollen and nectar to provision her nest. It is during this foraging process for pollen and nectar that she inadvertently pollinates the flowers she visits. In her chosen nesting tunnel she mixes pollen with nectar to form a pollen lump. When she has collected enough to feed a developing larva, she lays an egg on top of the pollen lump. She then closes off the nest cell with mud or plant material, or mixtures of mud and plant material. She will forage for pollen and nectar again, and create a new pollen lump, lay her egg and close off the nest cell to create an isolated chamber for each developing bee. She continues this pattern of foraging, provisioning the nest cell, laying an egg and closing off the next cell, until she has laid all of her eggs or until she dies.

2. Summer

Inside the chamber, the egg hatches into a larva or grub. It feeds on the pollen-nectar lump gathered by the female mason bee. As the larva eats, it grows. When fully grown, the larva enters a resting phase in which no feeding occurs. Then it spins a cocoon, changes into a pupa and, by the end of the summer, develops into an adult bee inside the cocoon. Growth of the developing bee largely depends on ambient temperature. Warmer temperatures will ensure that developing bees grow and mature. Cooler temperatures can halt growth and result in death before bee pupae develop into mature adult bees. This transition into adulthood occurs during the summer for *Osmia lignaria*. By September the pupae have developed into adult bees within their cocoons.

The female cocoons are usually larger than the male cocoons. Because mason bees are vulnerable to disease and predation during the transition from egg to adulthood, it is important that the development time is as short as possible. Research has shown that the effect of temperature on the rate of development is not the same in all phases (Kemp and Bosch, 2005). It was found that the optimum temperature for development from pre-pupa to pupa was 26°C (79° F). Additional time was needed if the temperature deviated from 26°C (79° F). In addition the rate of development from pupa to adult decreased with increasing temperatures. This means that cool temperatures in late spring to early summer slow down development from pre-pupa to pupa, and make these mason bees even more vulnerable to disease and predation. In addition, it means that very hot summers slow down the development of the pupa to adult. In other words, a warm spring and cooler summer are ideal for development time.

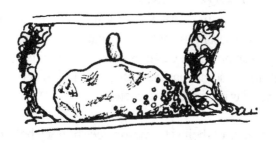

A view inside a nesting tunnel with an egg on top of a pollen-nectar lump surrounded by two mud walls.

3. Fall and winter

Both males and females remain in their cocoons until warm weather wakens them the following spring. Male cocoons are in the outermost cells, close to the entrance of the nest. Females are in the innermost cells, away from the entrance.

The winter period is hazardous for the bees. Insect-eating birds, such as woodpeckers, are known to wipe out whole populations of mason bees. Protection against predation of cocoons is a good part of mason bee management (Chapter 7)

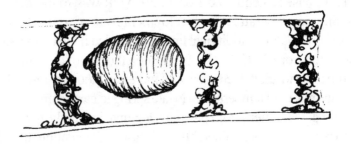

Adult mason bee in cocoon protected by mud partitions.

2.2 NEST REQUIREMENTS

Mason bees need a small dry nesting hole without a second entrance. They will nest in any potential nesting hole made of wood, masonry, plastic or metal. Cavities in any material is a potential nest site although some materials are preferred over others. For instance, wood is preferred over metal.

Probably because of their size, mason bees prefer a hole 7.5 mm (5/16th") in diameter. However, they nest in both smaller and larger nest cavities.

Mason bees use preexisting cavities for their nesting tunnels. Mason bees depend on burrowing-insects, birds or people to create nesting holes.

Mason bees nesting between cedar shakes on a house.

Mason bees nest in all kinds of cavities. In natural settings mason bees use holes created by creatures such as beetles and woodpeckers. Among the buildings in suburbia, mason bees have lots of opportunities to find available cavities. They use old nail holes and countersunk holes. Complicated structures such as key holes and electrical outlets are used when simpler nesting sites are not available. Mason bees nest in gaps between cedar shakes placed side-by-side. Another nesting spot is the gap left between a window and the window frame usually adjacent to weather stripping.

Mason bees may use keyholes as a nesting

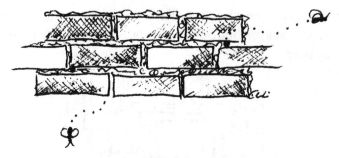

Mason bees nesting in a brick wall.

Opening the window destroys these carefully laid eggs. Other locations used for nesting are: a folded umbrella, honey bee comb and even vertical hanging metal wind chimes.

Structures such as homes and sheds provide many potential nesting sites for mason bees. For example, old brick and mortar buildings can provide nesting cavities in cracks and rotten masonry.

Unmanaged bee populations fluctuate widely in numbers as they go through pest and parasitic cycles. Learning how to keep parasitic populations low (Chapter 8) and bee populations high (Chapter 6) enables gardeners to have sufficient mason bees for pollinating their fruit trees every year. Buying or making a nest is the first step in keeping mason bees (Chapter 4).

2.3 FOOD FOR MASON BEES

Nectar and pollen, are the essential requirements for mason bees to feed themselves and provide food for their young. When food is abundant and available during the full length of the nesting period, mason bees will live longer and produce more offspring than if food in the form of flowers is scarce.

Take note of the dates when flowers are absent or scarce in your garden. If this period is the same as when mason bees fly, it is important to acquire new plants that fill this gap. Learn from other gardening friends and garden centres which blooms would be suitable for filling this period. Shop for flowers on a sunny day and watch for bees on the plants while you shop. Bees favour flowers rich in nectar and pollen. Just watch the bees and see what they like, then bring it home for your own garden!

If there are not enough flowers to provide food, your bees will search elsewhere. Many of my neighbours thank me for my abundant pollinators! Note that blooming flowers sprayed with insecticides will harm bees.

YOUR NOTES:

_____

_____

_____

_____

_____

_____

_____

_____

_____

_____

_____

Chapter 3

Getting Started with Mason Bees

Mason Bees require a nest, food, mud source and a good location to be successful in a garden.

Getting started with mason bees is as simple as fastening a specially-made mason bee nest to an outside wall of a building. A nest can be hung on a wall of a townhouse or an apartment building. You do not need a garden in the immediate vicinity to have mason bees. Mason bees have been successfully kept on the fourth floor of an apartment building, but higher locations may not work

as well for mason bees. In the spring, see if local bees find and occupy your nest. Attracting mason bees to your bee nest is similar to attracting wild birds to your nest. A bat house, bird house or a bee house is eventually filled with its respective occupants if it is a place that attracts them.

Mason Bees can be housed under either a Managed System or a Rustic System.

In a Managed System nests are managed to keep pests and parasites low so mason bees can thrive. Pests and parasites are removed each year so that healthy mason bees are present for pollination each year. There are quite a number of managed systems with varying levels of labour and cost (Chapter 4).

The Rustic System refers to a system of drilling blocks of wood and setting out additional blocks when nesting holes in previous wood blocks become full. The cost is relatively low although drilling holes is labour intensive. The biggest disadvantage is that when bee numbers are low, it is not possible to examine the nest and determine why there are so few bees. In nature, without any management, it is normal to see boom and bust cycles where there are few bees in some years and many bees in others.

3.1 NEST AND NEST PLACEMENT

Setting out a nest is a good way to encourage these excellent pollinators into your garden. Make or buy a nest. Make sure the nesting holes are at least 13-15 cm (5-6") deep (Chapter 4). This ensures that the correct proportion of males and females are produced. (Short nesting tunnels produce abundant males and few females, with the result that fewer females produce fewer offspring for next year's pollination.)

Place the nest in the sun and out of the wind and rain. East-facing is preferred, because mason bees start foraging when the early morning sun warms the bee inside the nesting tunnel. South-facing is also suitable because the sun warms the whole nest, although it is later in the day. Be cautious in setting nests in a southerly exposure. Excessive southerly heat from the direct sun may over heat and kill mason bee offspring.

Bees will start foraging earlier if the nest faces east and later if the nest faces south, west, or, even worse, north.

3.2 FINDING MASON BEES

If you set out nests and they remain empty, it's likely you don't have mason bees in your area. In that case, you can set out nests in other locations, then bring them home in the fall once the bees have taken up residence and are hibernating for the winter. Set out nests in early spring and wait for wild bees to lay their eggs. At the end of the summer, bring the nests back home. Harvesting bees from the wild by setting up nests, is unlikely to decimate wild bees because wild bees also nest in natural nest sites.

It is always best to set out nests in your home region. The bees in your home region are tolerant of the local climate and are more than likely to reproduce and multiply than mason bees from other regions.

Set out 5 to 10 nests in an area, approximately one nest every half kilometre (2/3 mile). Preferably set them out by permanent water sites, such as a creek or pond or where mud is available. Hang nests on old buildings, protected from wind and rain and preferably facing east. In cooler and more northerly climates, avoid hanging nests on posts and fences because nests are not protected from the wind as they are on a wall of a building. Increase air currents will create lower temperatures. Avoid trees because shade also lowers temperatures. Hot temperatures on the other hand are not tolerated by mason bees. Therefore in warmer climates, such as Alabama, Arizona and Utah, posts and trees are successful places for nest placement where bees need protection from extreme heat.

The abundance of mason bees varies between regions. In the Lower Mainland of British Columbia mason bees are found mainly in the suburbs, probably because of increased availability of garden flowers for food, and of nesting sites in man-made structures. In some locations such as Utah, mason bees are as common in the wild as they are in the suburbs.

3.3 PROVIDING FOOD

Once you have provided the bees with a sheltered nest, provide food in the form of flowers for a supply of both nectar and pollen. When food is abundant, it usually means more offspring. When a garden has few flowers present, food availability for bees is poor. In this case, bees may search for flowers in the surrounding areas but usually mason bees will go no further than two to three city lots to search for food. It is also important that flowers are available to mason bees throughout their foraging season. If flowers are both abundant and available continuously bees will produce lots of offspring for the following year.

3.4 PROVIDING A MUD SOURCE

Mason bees need mud to build partitions for nesting chambers. Mud is also used to create a plug at the end of the nesting tunnel. These mud partitions are used to prevent larva and pollen from drying out plus to keep predators and parasites out.

Usually, mason bees find a source of mud close to their nest. A mud source can also be created. Provide a mud source by digging a hole into the mineral soil of your garden. Create several rough areas along the vertical wall of the hole that has been dug into the dirt and mason bees will use these areas to begin excavating mud for their nests (Chapter 5.2).

YOUR NOTES:

_____

_____

_____

_____

_____

_____

_____

_____

_____

_____

Chapter 4
Choosing your Nest Type

Nests set out by gardeners provide nesting habitat for wild mason bees. Even if only a few wild mason bees use these nests, bee numbers are sure to increase in your garden.

This chapter describes the advantages and disadvantages of different nests created for mason bees. Mason bees will use all types

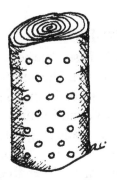

Holes drilled into log

and shapes of cavities. They cannot be too fussy because they may run out of time when flowers are in bloom and not be able to reproduce. Reproduction is the reason they go to work every day. A nest can have any number of nesting holes... from just 1 to 50 or more.

You can buy nests or make your own.

Nest materials and design depend on cost and labour. Usually if cost is low, the time spent making the nest is high. Making nests can be as simple as rolling paper tubes, drilling holes in wood or using a saw or router to make grooves in wood. Whatever nest type is chosen, bees will nest and pollinate flowers. The end result of pollination is larger fruit and more fruit.

You may want to try a variety of nests. All types of housing attract mason bees, some more than others. The fun part is seeing what the bees are doing. Is there bee activity in some nests and not in others? Which nest is more attractive to mason bees? Which nest is easiest to manage and clean? Which one produces the most cocoons?

4.1 KEY DIMENSIONS

The key dimensions of a mason bee nest are the tunnel inner diameter and the tunnel length. The inner diameter of the tunnel should be within the range of 7 mm - 8 mm (1/4"-3/8") or an average of 7.5 mm (5/16") and 13 cm -15 cm (5"-6") in length. Narrower, shorter tunnels result in a greater proportion of males. Usually, the front 8 cm -10 cm (3"- 4") of a tunnel is allotted to males and the rest is for females. Therefore if total tunnel depth is 8 cm -10 cm (3" - 4") bees will be mostly male which will lower the number of offspring.

4.2 NEST MATERIALS

Mason bees prefer some nest designs and materials over others. However, they do not always choose the nest types that are less prone to disease and pests. There are various rational ways for selecting and creating mason bee nests (Chapter 4.3). In all nest types, it is important to keep light out of the back of the nesting tunnel. If the nest has more than one opening mason bees will avoid this nest.

1. Wood
Mason bees will use wooden tunnels if the tunnels are kept clean and free of slivers. The nesting channels can be routered, cut with a table saw, or drilled into two adjoining pieces of wood. Alternately, you can drill holes into a block of wood (keeping in mind the measurements discussed throughout the chapter).

Avoid wood that has tree produced volatiles. This includes trees with a strong fragrance like cedar. New, green wood has a stronger scent than seasoned wood. Drilled cedar may have to "rest" for a number of years while the scent dissipates and before mason bees will move into the nest.

Wood for mason bee nests should be free of laminate preservatives and glue, all of which can be deadly to mason bee larvae. (Bees do not "naturally" avoid wood impregnated with toxins.) As well, plywood de-laminates in which the wood layers separate from each other. This makes plywood unsuitable for nests.

2. Paper/Cardboard
Paper can be hand-rolled into a tunnel nest. Most types of paper are suitable, with the exception of newspaper with oil-based print: the ink can be toxic to developing bees. Paper layered with plastic is not a good choice because it encourages mould growth. Mould that has grown over wintered cocoons does not appear to be detrimental to the bee inside the cocoon.

Commercial nesting tubes are made of cardboard and can be of

Quicklock nesting trays.

varying thickness and length. Some commercial tubes have a cardboard plug at one end of the tube, others have plastic end plugs. Some commercial tubes have a plastic liner on the inside.

3. Plastic
Beediverse® designed Quicklock Trays in 2005. Unlike wood, plastic trays don't warp, nor do they wear.

Beediverse® Royal House with Quicklock Nesting trays

Quicklock nests consist of two pieces of grooved plastic that can be locked together to form six nesting tunnels. These nesting trays can be stacked together and are ready to be set into a shelter. The inside of these tunnels are finished with a matte finish simulating a wood-like surface.

In the early development of mason bee housing, plastic drinking straws were tested as nesting tunnels. These are cheap and handy but unfortunately fewer cocoons were harvested from such nests. Bees appear to have difficulty entering plastic nesting tunnels. If plastic straws are inserted into a wooden block so that they are recessed, bees have less difficulty entering the nest. Nevertheless, productivity is low possibly because of the slippery surface of the plastic straws.

4. Corn
Quicklock Trays are available in a product line called Corn-material. This product is biodegradable under conditions met by commercial composters and contains more than 75% corn product. Trays made of Corn-material are most attractive to mason bees. Cleaning Corn-material trays is similar to the conventional plastic trays, quick and easy.

5. Styrofoam
Styrofoam nests are made for nesting alfalfa leafcutter bees and are used for commercial pollination of alfalfa crops. However, in my research, the mason bee *Osmia lignaria* avoids Styrofoam nesting tunnels whether in the form of trays or solid Styrofoam. In my studies bees always chose cardboard over styrofoam. Given only styrofoam, this material may be more successful than it appeared in this study.

6. Natural material
There are many plants with hollow stems that are used by bees. Reeds in marshes make useful nests. Some mason bee enthusiasts have harvested reeds and set them up in shelters for their mason bees. The following plants have also been used to house mason bees: Sorbus species, rush, cane and bamboo. Inert material such as concrete, rocks and snail shells are also used as nest structures for mason bees.

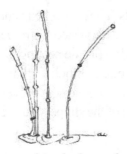

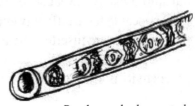

Reeds can be harvested
for use as mason bee
nesting tunnels.

A natural reed tube used
by mason bees as a nesting tunnel.
Small pollen lumps with eggs are
partitioned off from each other.

4.3 NEST TYPE

Nest types can be created from Tray Stacks, Tube Bundles and Drilled Blocks. All these nest types use grooved tunnels.

Grooved tunnels, properly cared for, will increase your bee population, largely because they can be dismantled and thoroughly cleaned. In the fall, mason bee cocoons can easily be removed and cleaned to avoid the buildup of parasites and nesting trays can be washed and prepared for the following spring. The removal of predators and parasites (Chapter 6) will increase your bee population more than is normally observed in nature.

1. Tray Stacks

Grooved channels are created by cutting into one side of a piece of wood using a 5/16th" router or simply with a table saw. Each piece of wood is called a TRAY. Trays are stacked on top of each other to form the nesting tunnels.

Each tray measures 15 cm (6") long, 9 cm (3 ½") wide and 1.9 cm (¾") deep. Wooden trays are placed on top of each other

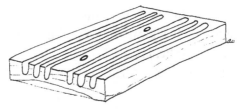

A single nesting tray with six routered nesting channels. The two holes are for bolting a set of trays together into a single nesting unit

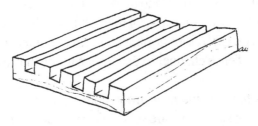

Tray cut with a table saw

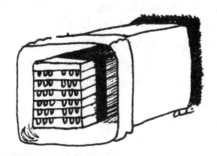

Trays can be bolted or taped together and housed to keep them out of the rain and wind

with the channels on top of the nesting tray. The last tray that is stacked on top is a blank piece of wood to form a lid over the nesting channels. Trays are tightly held together with bolts or tape to prevent predators from entering nesting tunnels from the sides. Routered trays are best made by routering wood to within 1 cm (1/2") from the end of a piece of wood creating blind holes (open only at one end). Three channels are clustered on one side and three channels are clustered on the other. If channels are routered along the entire length of the nesting tray, the stack of nesting trays will require a rear vertical board to create blind channels. An alternative method is to set trays inside and against the back of the housing.

Square channels can be cut into the piece of wood by passing it over the table saw with a Dado blade. Usually the cut is made into a full length of wood 1.8 m - 2.4 m (6'- 8'), which is then cut into 15 cm (6") pieces. It is important to remove splinters by sanding the insides of each channel so that bees will use the nest.

Quicklock trays are made from corn material and are available in small or large stacks that are set inside a wooden shelter and are suitable for setting outside. Each tray consists of two pieces of interlocking parts creating nesting holes. Trays can be stacked to any height and best secured with electrician's tape. The regular colour of these trays is light brown wood-like colour. Blue, yellow and green trays are also available. Coloured trays assist mason bees in finding their nesting tunnel when they return from foraging. The Quicklock nesting tray system is the easiest to maintain when compared to any other system.

*A Beediverse®
Highrise shelter
with Quicklock
nesting trays.*

2. Drilled Wood Blocks

Drilling holes in wood is an alternative nesting type. Check Chapter 4.1 for key dimensions. Since it is not possible to completely clean drilled holes, it is best to have a paper or cardboard insert into each drilled hole. Paper inserts can be made out of Kraft paper or newspaper and writing paper.

The number of holes drilled in one block of wood is important because if there are too many holes to choose from mason bees will have difficulty finding their tunnels. Even 25 nesting tunnels, set 1" apart will confuse the bees, and they will spend an inordinate amount of time searching for their tunnel. More time spent finding home means less time gathering nectar and pollen, pollinating flowers and laying eggs. Adding nest markers helps the bees find their way back home (Chapter 5.3).

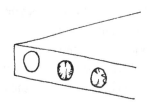

Holes drilled into small blocks of wood.

One variation of drilling holes into a large block of wood is to drill a few holes into the side of a board 2.5 cm x 2.5 cm x 15 cm (1" x 1" x 6"). A stack of these boards is inserted into a shelter.

In the fall, each block can be given to a neighbour. Be cautious in giving or receiving, nests or tubes filled with cocoons. This method of moving bees to another property may include moving parasitic mites and predators of mason bees. If you want to purchase bees, buy from a reputable supplier, someone who provides healthy and cleaned mason bee cocoons.

3. Tube Bundles

Nesting tubes can be made from cardboard or newspaper. Mason bees readily use nests with an internal diameter of 7.5 mm (5/16") and of any length. Commercially available cardboard tubes usually range from 9 cm (3 1/2") to 15 cm (6") long, although nests that are a minimum of 15 cm (6") in length are more productive. More females are produced in a 15 cm (6") tunnel than in a shorter tunnel. In contrast , a 9 mm (3 1/2") tunnel will produce fewer females and more male mason bees.

The walls of a commercially available cardboard tube are usually 0.5 mm (0.002") thick.

Nesting tubes can be made out of any paper, preferably Kraft paper without a plastic liner. The correct thickness is determined by whether the tube is going to be bundled, or individually inserted into a piece of wood with drilled holes. Wall thickness of a cardboard tube is critical. If cardboard tubes are inserted into drilled blocks of wood, it is important that the tube stays intact

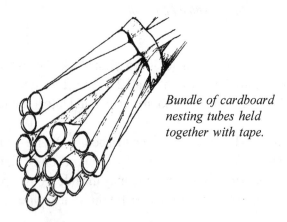

Bundle of cardboard nesting tubes held together with tape.

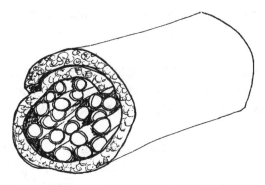

Tubes bundled into a piece of insulation

when removing them from the wood nesting tunnels in the fall for cleaning. Make sure that tubes do not fit too tightly into the wooden tunnels, because expanding wood and moist winter conditions will make it impossible to remove the tubes from the wood nesting tunnels.

For all types of tubes, wrap the bundle of tubes first with an elastic band and then with tape. If tubes are going to be bundled, it is beneficial if the ends of the tubes are uneven in length as they can find their nests more readily. Surround the bundle with insulation

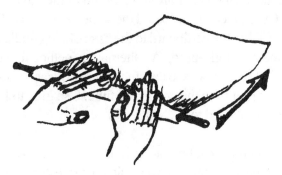

Making hand rolled paper tubes

such as foam or newspaper. Insulated tubes are less likely to over-heat and kill developing bees.

Hand rolling paper into nesting tubes and bundling them together is a simple way of making nesting tubes. This method was intro-duced to the author by Research Associate, Wei Shuge.

You will need paper cut into strips and a dowel for rolling the paper around it. Cut 18 cm x 5 cm (7" x 2") segments of plain white paper and longer pieces of newspaper. Use a 7.9 mm (5/16") diameter, 25 cm (10") long metal or wooden dowel. Place white paper on top of newspaper to avoid ink coming in direct contact with larva and roll both types of paper around the dowel to create a 18 cm (7") long tube. Securing roll with liquid glue makes it difficult to unroll tubes when harvesting cocoons in the fall. Ease the dowel out of the centre of the paper tube after tube has been taped.

When 10, 20 or more tubes have been made, bundle these tubes with an elastic band and then wrap the bundle with packing tape, leaving the holes exposed. Next, cut a 15 cm x 15 cm (6" x 6") piece of paper. Coat one side of the piece of paper with glue and wrap glue coated paper around one end of the bundle of rolled tubes. Cut a piece of cardboard the same diameter as the bundle of tubes. Glue the cardboard to the base of the bundle to ensure the tunnels are light-proof. An alternative method for closing off one end of each tube is to cut small pieces of cardboard that are slightly larger than the end of the tube. Use glue to fasten one piece of cardboard to each tube.

Wrap the bundle of tubes into a piece of insulation. A roll of recy-cled newspaper or cotton type insulation protects bees from over-heating during summer temperatures.

Paper tubes or inserts are available for cardboard nesting tubes and drilled nesting holes inside the wood. Place inserts inside cardboard tubes or wooden blocks with drilled holes. Inserts are removed from their housing together with bee cocoons in the fall. Inserts are disposable and are destroyed once the cocoons have been removed. In the spring, new inserts give the bees clean nests. Examine nests for mites. If mites are present, nests will require cleaning prior to inserting the new paper inserts.

As an alternative to commercially available inserts, hand rolled Kraft paper tubes may be inserted into blocks of wood with drilled holes. Roll a suitable piece of Kraft paper around a dowel and insert dowel with the rolled Kraft paper into each hole. When paper is completely in drilled hole, remove dowel. This is repeated until all tunnels are filled with an insert.

YOUR NOTES:

_____

_____

_____

_____

_____

_____

_____

_____

_____

_____

_____

Chapter 5
Preparing Nests for Spring

The most important step in starting with mason bees is to have a nest ready for bees in the spring. Mason bees search for nesting sites in early spring, several weeks prior to spring bloom. For coastal British Columbia in Canada, set nests out in February and earlier in southern regions. On the west coast of British Columbia, Washington and Oregon, natural emergence occurs when the flowering shrub Japanese Andromeda (*Pieris japonica*) begins to bloom, usually after willow and before dandelion. Choose the nest that is best suited to the needs of your garden, and be sure it is clean and free of pests or mould.

5.1 NEST PLACEMENT

Place nests close to the plants you want pollinated. This is important because the closer the nest is to the flowers, the more time is spent foraging for food and providing for their young. The best location for bees to produce many offspring is not always obvious because wind and shade make nests less attractive to mason bees. When nest location, weather and food are optimal, mason bees can increase to six times the original mason bee population. Similarly, time available for foraging can be increased by making mud available close to the nest and by placing nests close to flowers.

1. Location

The best nest location is above ground, in the sun, protected from wind and rain and facing east. Bees need the sun to warm up before they are able to fly. When a nest faces east, the bees have a good chance of warming up in the early morning sun. This is why bees prefer east facing nests. When nests face north, mason bees do not begin foraging until mid-morning, when ambient temperatures warm to 16°C (61°F). North facing nests are cooler in the early morning and these mason bees will have less foraging time than mason bees nesting in easterly facing nests.

Avoid placing nests on fence posts or on a fence because air movement around the fence has a cooling effect on the nest. Also, avoid placing a nest in the shade, on the north side of a building, or in the shade of a tree. In contrast, excessive heat in the hotter, drier climates such as southern states of the U.S. and the intermontane regions of the northwest can also be detrimental to the bees. Therefore, nest placement that has noon or afternoon shade in hot climates is usually desirable.

2. Height

Mount the nest about 1.5 m (5') high on a wall of a shed or a house. Bees will use nests placed at or near the ground or higher than 1.8 m (6'). Five feet is a good height for observing your bees flying to and from the nest.

3. Distance

When bees are close to their food source, more flowers are visited and pollinated, and less time is spent travelling to and from their home. Place the nest as close as possible to flowering trees and other flowers. Within a distance of 70 m (225') works well. The greater distances that bees have to travel between their nest and foraging patch, the less pollinating is completed. Nest placement is best where food source and orchard blooms that need pollination are close to the nest.

4. Moving nests

Once the nest is placed out in early spring, **do not move it until midsummer**. Spring is the critical foraging time for mason bees. Returning females can become lost if their nest is moved while they are out foraging. Once separated from their food supply, larvae starve and die. Jarring or dropping the nest in the summer can damage the developing bees inside the cocoon.

If you must move the nest before fall, take the entire unit, keeping the tunnels horizontal and being careful not to jar it. If you do happen to jar the nest, tip it and place it in the new location with the entrance skyward (and the tunnels vertical). That way any dislodged larvae will fall back onto their food mass and continue to eat and grow.

The time to move a nest is in September when it should be cleaned and transferred to a non-heated garage or shed, to wait until spring.

5.2 MUD AVAILABILITY

Mason bees need mud to plug up their nests and to construct partitions between each egg laid on its pollen provisions. These partitions seal the newly laid eggs from predators and parasites. Because the mason bees' time of gathering mud is time away from foraging (i.e. pollinating) it is best to provide a mud source nearby. When mud is close-by bees will spend more time foraging and pollinating fruit blossoms instead of travelling long distances to a mud source.

To provide bees with a mud source, dig a hole 30 cm x 60 cm x

30 cm (1' x 2' x 1') within 15 m (50') of the nest. Dig through the darker organic humus layer into the lighter, clay-like mineral soil. If soil is dry, add water. You may see bees collecting their muddy load, and you may see them flying back to the nest. You'll notice that bees dig into the side wall of the dug hole to gather their mud.

Watching mason bees collect mud from a hole dug into to the mineral layer of the soil.

If your soil consists of gravel and sand, you can still provide a mud hole for your bees. Dig a hole 60 cm x 60 cm x 30 cm (2' x 2' x 1'). Place a plastic ice cream bucket at the bottom of the hole. Fill the bucket to overflowing with mud. Pour water into the bucket on a weekly basis. Water from the bottom of the bucket will permeate up through the mud to moisten it. Bees only collect mud with enough moisture to mold mud into tunnel plugs.

5.3 NEST RECOGNITION

When a bee begins foraging for pollen and nectar, she uses cues to orient herself back to the nest entrance. In early spring you may see mason bees going 'in' and 'out' of many nesting holes within seconds. These bees are confused by the lack of markers. Decorating the front of a nest makes it easier for bees to find their own nesting tunnel. A nest can be decorated with simple designs using one or two colours. Note though, that bees are easily confused by too many colours or complex designs or letters. If there are too many markers, bees fly in and out of several nesting tunnels, just as if there are no markers. You may even see them being thrown out of a neighbour's nesting tunnel. They waste valuable pollinating time flying around looking for their nesting tunnel. Yellow, mauve, pink and blue are good colours for marking the front of each nest. Simple designs painted across the face of a nest help a bee find her way back to her nesting tunnel. Some simple designs are shown here.

L   I   V   X   O

Simple designs on the face of a nest improve
mason bee nest tunnel orientation

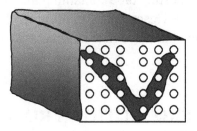

These simple designs must be large enough for the bees to discover them as they approach the nest. For example, make each part of the 'V' 10-15 cm (4-6") long depending on the size of the nest. Thickness of each line can be around 1 cm (0.4").

Tube bundles should also be decorated with paint. Paint the end of half the tubes with one or two colours. The unevenness of tubes in a tube bundle is an additional cue for bees. Too many markers, designs, or colours can also confuse the bee and disorient her so she has a difficult time finding her nest.

Recent research on visible nest cues that may be used by *Osmia lignaria,* demonstrates that contrast colours of black/black, black/grey and black/blue, set at 1 cm, 2 cm and 6 cm had an effect on the confusion level of the bee. Confusion by bees increased with increasing contrast and depth (Guedot et. al., 2007).

It is also apparent that mason bees use chemical cues for recognizing their own nesting tunnels. In 2003, Bob Chappel (Victoria, B.C. Canada) filmed a female depositing a clear droplet of fluid from her abdomen around the inner wall of the nesting tube in a spiral fashion along the length of the hole. She deposited the fluid while rotating around inside the nesting tunnel (DVD, Pollination with Mason Bees by BeeDiverse®).

In 2006, this fluid was analysed and tested (Pitt Singer et. al. 2008). When a nesting tube containing the fluid deposits was moved, the females reacted as if "confused by the lack of scent". This behaviour indicates that this material is used for nest recognition.

5.4 NEST PROTECTION FROM PREDATORS

Predation of nests occurs by a variety of animals.

Nest protection from squirrels can be solved by setting nest inside a sturdy box with one hole in the front allowing bees to exit and enter.

Ants can be a problem because ants also eat the pollen and honey. One solution is an ant trap with Tanglefoot®, a commercially available sticky substance that does not dry out when set outside. Dr. J.H. Cane, Research Entomologist, USDA, Logan, Utah, has created a unique method of keeping ants off bee nests. A tennis ball is cut in half and punctured in the center of both halves. The lower half of the tennis ball is filled with Tanglefoot® and the upper part of the tennis ball acts to protect the container of Tanglefoot® from filling up with rain water. Both halves of the tennis ball are skewered onto a conduit pipe. Cane also suggests that an area can be tested for potential ant problems by setting out several baits made from honey smeared onto paper. In regions with a high ant populations, ants find honey within minutes.

Predation of mason bee nests by bears can usually be prevented by setting up an electric fence where nests are set up.

Woodpeckers feed on mason bees in the summer, when young bee larvae have grown into adults. They do not eat mason bee grubs present in the nest in the spring, nor do they eat flying bees. Protect the young adult bees in the nest by crimping chicken wire over the front of each nest. Do not place chicken wire in front of nests during the spring because chicken wire will make nests less attractive for nesting. Woodpeckers have a long tongue, so make sure there is a generous gap between the chicken wire and the face of the nest(s). Use firm chicken wire with 2.5 cm (1") diameter holes around the front of the nest. Place the chicken wire 5-8 cm (2-3") away from the face of the nest so that the long tongue of woodpeckers cannot reach the mason bees.

5.5 NEST PROTECTION FROM RAIN

Nests with young developing bees can easily be destroyed by rain. Protect the nesting unit with good roofing material like cedar.

The whole nest can also be placed in either a ceramic or plastic shelter that is big enough to protect the whole nest. Attaching a square plastic bucket (on its side) on a wall works well. If the container is white or light in colour, paint it black or dark brown because bees appear to prefer darker shelters. Bees tend to prefer nests that are placed in sunny locations with shaded interiors. Shaded nests can be created with an overhanging roof. Drill a few holes at the bottom of a plastic bucket or similar container to allow ventilation and avoid excessive humidity.

5.6 NEST PROTECTION FROM PARASITIC WASPS

Protecting cocoons from wasp parasitism during the summer months is a good part of mason bee management. You can use a net bag like the ones sold by Beediverse®. This netting material protects your mason bee cocoons from pests and parasitic wasps and other flying insects while they mature inside the nest. A finely woven mesh bag (60 cm x 60 cm, 24" x 24") is efficient, lightweight and easy to use. The netting bag can be draped over one or more mason bee nests. Air can enter and exit easily through the mesh, helping to keep the cocoons healthy.

Draping the nests with netting keeps parasitic wasps out and traps wasps already in the nest. While wasps trapped in the nest can still develop, they can't exit the nest to mate - providing some control over the parasites' population. You can also spray the trapped wasps- inside the net bag with a mild soapy solution or even a light application of hairspray (Chapter 10.4). Both sprays kill parasitic wasps.

5.7 NUMBER OF NESTING TUNNELS

The number of empty nesting tunnels required in the spring depends on the number of mason bees. If you have seen mason bees in the garden, set out several nests. Place nests in different locations to increase the chance that wild mason bees will find one of them. More mason bees are usually found in gardens with old buildings, sheds, dead snags in trees and/or around houses with cedar shingles, siding and roofs.

Once a nesting season is complete and cocoons have been harvested and counted in the fall, a better assessment can be made of how many nests are needed for the following spring.

The number of nests required for the following spring can be calculated with some certainty. For example, 100 cocoons harvested in the fall, have about 45 females (normally about six males are produced for every five females). On average, each female mason bee uses one nesting tunnel. Therefore, about 45 nesting tunnels are needed for 100 cocoons. This calculation may seem simplistic, but on average, in a good year, one female only requires one nesting hole. If a female lays all her eggs (about 30), she will need about five nesting tunnels. However, birds, spiders, food supply and weather conditions all take their toll. In addition, most female bees only live about three weeks. The 100 cocoons will mature into adult mason bees next spring. They in turn, can be expected to produce 250-100 offspring. The survival rate of those offspring is between 30-50%. So you should anticipate 100-300 cocoons next year and provide nests accordingly.

No matter what system of nests is in place, a common question is, "How many mason bees do I need for my backyard?" If there are fruit trees present, decide whether you are satisfied with the amount of fruit produced over 3-4 years. If the amount of fruit is satisfactory, you have a sufficient number of mason bees for your garden.

YOUR NOTES:

Chapter 6
Fall Harvest and Cleaning

Fall and winter management includes harvesting, cleaning cocoons and cleaning the nest. The purpose of cleaning cocoons and nests is to maintain bee numbers sufficient to pollinate the fruit trees in the garden. The number of bees can be increased or at least maintained by keeping predators and parasites numbers low. If bees and their nests are not cleaned on a yearly basis, bee numbers go up and down over the years, as in all natural cycles. The result is that pollination and fruit harvest will also go up and down. It is recommended that you clean both cocoons and nests each fall.

6.1 HARVESTING COCOONS

Harvesting cocoons is a simple task when nests can easily be opened as in the nesting tray system. For example, harvesting is more time consuming if the nest has to be opened by tearing open cardboard tubes. First prepare work site with a large piece of newspaper to contain debris and fragments of dried mud removed from nests. Methods for removing and harvesting cocoons varies depending on the nest type.

Nesting trays are easily disassembled.

1. Nesting trays

Open nesting trays by removing bolts, tape or other fasteners.
Then examine the trays one at a time.

*Mason bee cocoons are easily harvested from
nesting trays by separating one tray from another.*

Remove cocoons from each tray by using a tool that slides underneath the cocoons so they can be lifted up and out of the nesting
channel. A "sharpened" coffee stir stick may be used.

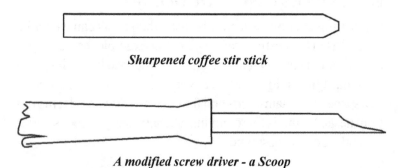

Sharpened coffee stir stick

A modified screw driver - a Scoop

A 'scoop' is a specialized tool designed and first produced by
Randy L. Person W.A. United States. It can be made by grinding

a sharp curve into a tool such as a screwdriver. Gently move scoop along base of nesting tunnel underneath cocoons and lever them out of each nesting tray. If the tray is held vertically and supported on a counter, the tool can be pushed vertically down underneath cocoons lined up in each nesting channel. If a section of the nesting tunnel is filled with mites (reddish sand-like material), remove the cocoons while leaving a majority of mites behind in the nesting tunnel. Separating mites from cocoons at this stage will significantly reduce your washing time. After cocoons have been harvested, set trays aside for cleaning.

2. Paper or Cardboard tubes

Cocoons are removed from cardboard tubes by making a small cut in one end of the tube with a sharp utility knife. Then, the cardboard is unravelled from this point and cocoons are placed into a container. It is more time consuming to harvest cocoons from cardboard or paper nests than from nesting trays.

Making a small cut into one end of the cardboard tube.

Cutting cardboard

Unravelling

3. Blocks of Wood with drilled holes

Blocks of wood with drilled holes cannot be cleaned without destroying nest contents. In spring, nest contents would consist of developing young bees or mature mason bees inside their cocoons that have not emerged. Exact nest contents would depend on whether mason bees have started nesting and laying eggs. Because cocoons are fragile and cannot withstand any kind of pressure onto the cocoons, pushing a rod through a nesting tunnel will destroy the bees inside their cocoons. Even if tunnels can be opened at the back of the nest, cocoons are too fragile to be pushed through the nesting tunnel. It is not possible to harvest cocoons from drilled holes in solid blocks of wood unless these nests have paper or cardboard liners. Neither is it possible to fully clean debris out of nesting tunnel, while bees are using nests, without destroying bees inside the nesting tunnel. At no time in the winter, spring or summer is the nesting tunnel free of bees because nesting time and emergence time overlaps. For example, some bees start nesting early in the spring while others are emerging from their nests. Thus, cleaning the nest with a drill in the spring is counter productive because drilling will destroy overwintering bees, newly laid eggs and developing young.

4. Recycling Blocks of Wood with Drilled Holes

Because mason bee cocoons cannot be removed from wooden blocks, another method can be successful depending on the mite level inside nests. This system alternates the use of nest blocks between years.

In the first year a nest is set out and becomes occupied with hibernating bees. Prior to bee emergence in the following spring, place this nest block upright in a cardboard box and tape cardboard box shut. Ensure that all gaps in box are taped to stop light from entering. In other words make the box lightproof. Then create one exit hole for the bees at the base of the box by pushing a pencil through one side wall. Place the box in a carport or shed with open doors, where it is relatively dark. Set out a second nest block on the wall of the building ready for emerging bees. In the spring bees will emerge from the nest, go to the only source of light coming into the box and exit from this box. After exiting bees will look for another clean nest. The presence of a clean nest on a wall nearby, will make it less likely for bees to go back inside the cardboard box and nest inside the used and dirty nest block. If cardboard box containing the nest block is placed in the sunshine, some mason bees may learn to recognize this small entrance to their natal nest, enter and use the nests inside the box. Bees generally do not go back into the old nest inside the cardboard box when placed in a dark location and if other nests are readily available for emerging females. Once the bees have finished flying, you can clean the old one.

6.2 CLEANING NESTS
1. Why Clean Nests
There are many reasons for cleaning nests and cocoons. Bees prefer clean nests, and clean nests result in plentiful and healthy bees. After just one year, every nest contains old cocoon debris, mud particles, dead bees and parasites, including pollen feeding mites. Mites that remain on the outside of cocoons will migrate to the bee as it emerges. Dick Scarth (Vancouver, B. C. Canada) observed that as soon as an emerging bee chews a small hole through the cocoon, mites on the outside of the cocoon immediately migrate into the cocoon and attach themselves to the emerging bee. Pollen-feeding mites will compete with the mason bee larvae for the food-source, and will load down the newly emerged mason bee and thus impede its flight (Chapter 8.1). If the bee is loaded down with an excessive numbers of mites, the extra weight

from mites can prevent bees from flying altogether. Clearly, hygiene is an important part of managing your mason bees.

2. When to Clean Nests
The best time to clean mason bee nests is from October to December because adult mason bees are fully developed and sturdy enough inside their cocoons to withstand the removal and cleaning process.

Bees, between the months of September and December, do not emerge from their cocoons if held at room temperature. During these months, bees go through physiological changes required for their development. After December bees can emerge in less than 24 hours if left at room temperature. Therefore, after December

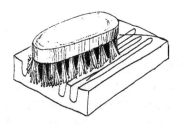

Scrubbing brush for cleaning trays.

you must limit the number of hours that cocoons are kept at room temperature. Once bees emerge, they cannot be kept alive until spring. The time taken to process cocoons can be lengthened if completed under cooler temperatures. By February in particular, male mason bees require only about two to three hours at 20°C (68°F) to emerge from their cocoons. Thus, it is safest to clean the nests and prepare the bees for storage in October.

3. How to Clean Nests

After cocoons have been removed from their nesting tunnels, soak nesting trays in water to soften any adhering mud. Once mud has softened use a scrubbing brush with lots of water to remove mud and debris. Then, soak trays for about two minutes in 0.05% bleach solution [add 15 ml (1 tbs) of bleach (5% sodium hypochlorite) to 4 L (4 quarts or 16 cups of warm water)] to kill adhering bacteria and fungi. Rinse well with running water to remove all traces of bleach. Check that the acrid odour of bleach is no longer present.

6.3 CLEANING COCOONS

Cleaning cocoons involves washing, screening and drying and can prevent the spread of disease. It can also keep young mason bees free of disease for the following year.

1. Why Clean Cocoons

The advantage of cleaning cocoons is that cocoons can be inspected and the contents of each cocoon can be determined. Diseased and parasitized cocoons can be identified and removed. Removal of suspect cocoons can prevent escalating problems with disease and parasites. After cleaning, good quality cocoons should be separated from parasitised cocoons and be prepared for the following spring. Be sure to clean all cocoons, including the cocoons that are not covered in mites, because it increases the chance of identifying wasp parasites and other diseased cocoons.

2. When to Clean Cocoons

The best time to clean cocoons is when they are harvested in the fall. At harvest time, the cocoons are placed into large containers such as a 4 L (4 quarts or 8 cups) ice cream bucket. Buckets are filled with water. If this cannot be done at harvest time, lay cocoons in a bucket without the water, and store dry until cleaning process can be completed. Avoid crushing the cocoons by minimizing the number of layers. Do not have cocoons deeper than 5-8 cm (2-3"). They are normally covered in faecal pellets and dried mud from nest partitions. Both the mud and faecal pellets can easily be washed off. If you live in coastal regions, expect some cocoons to be covered with mites.

3. Washing cocoons
Cocoons can be cleaned in water because cocoons are buoyant and repel water. Washing cocoons is relatively easy, although

Cocoons are first placed into a bowl of tepid water.

somewhat time consuming. Never use soap or detergent because soap-like materials will go through a cocoon and kill the bee inside.

4. Removing Mud from Cocoons
First, remove mud by placing cocoons in a 4 L (4 quarts or 8 cups) bowl of tepid water. Wet the cocoons by gently rolling and moving them through the water in a bowl. At first, mud falls away from cocoons and sinks to the bottom of container. Let cocoons sit for 20-30 minutes in the water, occasionally stirring them to

A sieve is used to rinse cocoons free of debris and mites.

ensure mud on cocoons is well wetted down. Take cocoons out of the bowl being careful not to loose cocoons down the drain. Throw out the dirty water and throw the debris of mud and sand into the garbage. Some of these debris will be clumps of mites.

5. Removing mites from cocoons

The majority of mites are removed from cocoons by washing. Set up a washing system so that cold water runs in at one side of a bowl, past cocoons contained inside the sieve, and out of bowl at opposite side. Set bowel at an angle to achieve this effect. Gently agitate the cocoons with your hand to make sure cocoons and mud are well wetted for easier removal of both mud and mites. Set the water level in bowl below edge of sieve. This will prevent cocoons from going into the sink. Look closely at the surface of the water for a decreasing number of reddish mites that float away during the washing process.

Prepare a bleach solution (approximately 0.05%) by adding 15 ml (1 tbs) of bleach (5% sodium hypochlorite) to 4 L (4 quarts or 16 cups) of warm water. Lower the sieve containing the cocoons into the bowl. Submerging the cocoons into bleach will further wet them down and remove additional debris and remove any mould.

After 10 minutes of this washing process, gently circulate cocoons in water and leave them under running tepid water for another 5 minutes to rinse of the chlorine.

6. Drying Cocoons

Drain cocoons using the sieve and gently turn cocoons onto several layers of white paper towel. Place an extra layer of white paper towels over the cocoons and wet it down. After an hour, remove towelling from cocoons. Look for tiny orange spots on the paper. If there are less than 10 mites per 6.5 cm^2 (10 per square inch), the first washing was a success. Additional mites can be removed by screening (Chapter 6.3.1) and immobilization (Chapter 10.4).

*A white paper towel is used to assess
mite levels on cocoons after cocoons have been washed.*

7. Screening Cocoons for Mites

Screening cocoons to remove adhering mites is best done on metal window screen. The screen can be stapled onto a wooden frame. The depth of the screening tray should be a minimum of 5 cm (2") so that cocoons are safely contained onto the tray. A metal kitchen sieve is also suitable for screening cocoons, although the volume that can be done effectively is less than on a flat tray. The friction of cocoons rolling over the metal screen will dislodge most mites.

When cocoons are dry (about an hour at room temperature after the washing process), set a handful of cocoons onto the metal screening tray. Move tray sideways and cocoons will roll around. Roll cocoons around for 30-60 seconds. When screening, do it over newspaper, bathtub or sink, so all mites can safely be cleaned away.

When the majority of mites have been removed, cocoons can be placed on another set of clean paper towels for any additional drying. After an hour, most water will have evaporated. Cocoons can now be sorted to determine whether they are filled with mason bees or parasitic wasps.

6.4 REMOVING PARASITIZED COCOONS

Because parasites can devastate populations of mason bees, parasites must be kept to a minimum. The easiest way to control parasitism is to remove cocoons containing parasites. Cocoons with adult mason bees are firm to touch and dark-grey in colour. Cocoons that are parasitised by wasps are usually lighter in colour, empty in appearance, less firm and more 'crispy' to touch than cocoons containing a mason bee.

1. Inspecting cocoons using a light box or flashlight.

Another method for inspecting cocoons for parasitism is to place them over a light — just like candling eggs (J. Sadowski, Burnaby, B.C., Canada). A flashlight with a rim around the light is most suited for candling cocoons. Place enough cocoons on the face of the light to cover the entire glass plate. Set this up in a dark room and turn flash light on. Parasitised cocoons are transparent whereas mason bees can be seen in the fetal position inside the cocoon.

Caution: do not leave cocoons over light for longer than a minute, because heat produced by the light may dry out and kill the bee. For examining large number of cocoons a light table can be constructed. However, to obtain enough light to see the inside of the cocoons and the least amount of light so as not to get blinded by the light, is a difficult balance to reach.

2. What to do with questionable cocoons

Questionable cocoons may be thrown out immediately, but there is a good chance that questionable cocoons may in fact contain live mason bees. An alternative to throwing questionable cocoons out, is to keep these cocoons in a ventilated container. Store the container with questionable cocoons in a cool place such as an unheated garage.

In the spring, bring the container inside and keep it at room temperature. Insects will eventually develop and emerge. If mason bees emerge from some cocoons, cool them down in the fridge for 30 minutes and release them outside. If the cocoons contain tiny parasitic wasps (about 1/10th the size of a mosquito) cool them down in the fridge for 30 minutes then destroy the temporarily immobilized wasps.

3. What to do with parasitized cocoons
Infested cocoon contains about 60 developing wasps, usually in the pre-pupae (undeveloped wasp) stage. If a parasitized cocoon escapes detection, adult wasps emerge in May and June. These wasps enter the bees' nest and parasitize developing bees. Every three weeks, adult wasps emerge and parasitize additional developing bees. This process continues until the end of the warm weather or until bee nests are taken down and placed into storage. Every new wasp can parasitize at least one mason bee cocoon. Twenty wasps could infest 20 cocoons. The resultant 200 wasps (20 wasps x 10 offspring per wasp) could infest an additional 200 mason bee cocoons. Removing parasitized cocoons at harvest time prevents a significant level of re-infestation and loss of cocoons.

Container with cocoons suspected of being infested with parasitic wasps.

Chapter 7
Winter Storage and Spring Release

Even after cocoons are harvested and cleaned, mason bee cocoons are vulnerable to predators and parasites. Therefore correct storage is necessary to protect and ensure cocoon survival through the winter months. When mason bees are hibernating inside their cocoons, they require physical protection and protection against extreme temperatures.

7.1 PHYSICAL STORAGE

Often simple cardboard boxes alone do not provide sufficient protection for mason bees over the winter.

1. Container for Cocoons Storage

First, store cleaned and air dried mason bee cocoons in a cardboard box, cushioned with toilet tissue. Place the cardboard box inside a metal box such as a coffee can. This prevents mice from eating the cocoons. Puncture two or three holes into the metal container for air circulation. Label the container MASON BEES so it does not accidentally become trash.

Examples of containers for storing mason bee cocoons.

2. Mould

Mould grows on cocoons if they are stored wet or if cocoons are stored under damp conditions. The holes in a metal container usually prevent moulds from growing over the cocoons. Mould growth is particularly vigorous if pollen is left on the cocoons' surface.

However, it does not appear that this mould is detrimental to mason bees inside the cocoons. If you choose to remove the mould, rinse the cocoons in a bath of mild chlorine solution. Use 0.05% bleach [add 15 ml (1 tbs) of bleach (5% sodium hypochlorite) to 4 L (4 quarts or 16 cups of cool tap water) and rinse for about five minutes. Rinse several times with water to remove bleach from cocoons and air-dry. Cocoons usually dry in about an hour, especially if the paper towelling is changed several times.

Do not attempt to wash the nest or cocoons once the bees begin to emerge. Getting wet can be lethal to the newly emerged bees.

3. Storage temperature

Under natural conditions, mason bees may undergo several weeks of below freezing temperatures. Under west-coast conditions freezing temperatures are moderated by coastal conditions. Nevertheless in British Columbia, Canada and northwestern United States below zero temperatures can extend for several weeks. For cocoon storage, extended extreme winter temperatures may be detrimental to survival. Natural extreme periods of cold temperatures can moderated by setting containers of mason bees in a shed or unheated garage.

The research on emergence time and survival of the European mason bees, *Osmia cornuta,* related to overwintering temperatures provided some interesting facts (Bosch and Kemp, 2004). Results indicate that mason bees need cold winter temperatures to emerge successfully. Emergence was generally unsuccessful with less than 30 days of winter temperatures. Winter duration of 90-150 days was optimum for maximum survival and longevity. Another interesting fact is that larger bees are more likely to survive the winter and live longer than smaller bees.

4. Refrigeration
Refrigerating cocoons can delay emergence and ensures the bees don't emerge earlier than blooms. In the Pacific Northwest, cocoons are typically transferred from their outdoor winter storage to a fridge in early February — just before the onset of warmer spring temperatures.

Correct temperature and humidity are critical for mason bees under refrigeration. Set the fridge at 2-4°C (36-39°F) and control for 60% or more humidity. Above 4°C (39°F), males will emerge while inside the fridge. Below 2°C (36°F) the bees may freeze. Freezing during early spring may cause mortality of mason bees. Home fridges are not made for scientific purposes so temperature and humidity will fluctuate. Check and adjust regularly.

5. Manual defrost Fridge vs Frost-free Fridges
Manual defrost fridges are more humid than the frost free fridges. This makes them the preferred environment for mason bee cocoons that require 60%+ humidity. As a precaution, place one or

two open containers of water in the fridge to increase the humidity and keep the environment moist.

Most home owners have frost-free fridges that run very dry. Low humidity can dehydrate and subsequently kill the bees. So you have to make a high humidity environment within the fridge. Place a damp paper towel inside an airtight container. Place cocoons on a lid from a jar for example, and place this lid directly adjacent or partly on top of damp paper towel, inside the airtight container.

7.2 SETTING OUT COCOONS IN SPRING

You can control when the bees emerge by extending refrigerated storage time, or shortening storage times and managing the temperature and humidity. If fruit trees bloom at the same time as the bees emerge, simply set out cocoons and let nature take its course. If fruit trees bloom before bees emerge, bees can be warmed and emerged earlier. If bloom, such as blueberry bloom, appears later cocoons can be cooled to delay emergence.

1. Field Structures for Nests

Protecting a mason bee nests from rain and having the nests in a warm place is a challenge in fields with crops because during the spring there are periods of rain or cool weather. In addition, setting nests out into fields or orchards, is often a challenge because there are limited number of sheds for attaching the nests. Thus, some form of structure is needed so that nests can be kept above the ground, away from the rain-splash zone and protected from the rain.

Simple boxes can be set up and are best if they are at least 60 cm (24") off the ground. An overhanging roof is another big plus to keep rain off nests.

Another structure we have tested successfully is the yurt. The idea came from shelters used in the alfalfa leafcutter bee industry. It is a six or eight sided structure framed out of wood or rebar and covered with tarp material. A hole in the center of the roof, 25-38 cm (10-15") diameter, allows excess heat to escape. Temperatures inside the yurt are warm during the day and temperatures on very hot days are moderated. Warm temperatures allow mason bees to emerge faster, start nesting sooner and allow their offspring to develop sooner.

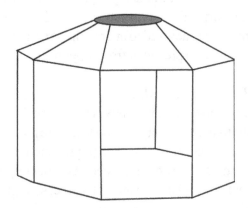

Eight sided structure named the yurt.

We originally created a frame of wood, then irrigation pipe and now we are testing a welded rebar unit and a unit made from sturdier wood frame. The interior diameter is about eight feet in diameter. The walls of the eight or six-sided structure are vertical and about 1.5 m (6') tall . The roof is open in the center and slopes from the top of the hexagonal wall to the center opening. The center opening with the roof with its 10-15" diameter hole allows excess heat to escape.

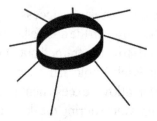

Eight spokes come from the center hole of the yurt.

The frame is draped with tarp material. Tarps are either blue, orange or white. The tarp is extended about one foot longer than the walls, so that dirt can be added and protect wind from getting underneath the tarp. If the mason bee yurt is less than about a meter in diameter it is advised to anchor the structure with rebar feet and or tie downs to prevent it from toppling over in the wind.

2. Natural Emergence of Bees
Allowing the bees to emerge naturally is fairly straightforward but some precautions are necessary. Cocoons require protection from birds and other predators that feed on unprotected cocoons and bees that are emerging from their cocoons. Cocoons also need protection from rain and direct sunlight. Direct sunlight may dry out or cook the bee inside the cocoon.

3. Emerging Boxes
If there are less than 50 cocoons, place them in a small box with a lid. Cut a small hole into the box so that bees can exit the box. Place this box adjacent to the mason bee nest. If the box is small, it can be tucked into or above the mason bee nest. Do not use boxes that have been used for pesticides, or petroleum products. These residues may kill the newly emerged mason bees. Even washing boxes may not completely remove these deadly toxins. If you harvest thousands of cocoons, distribute them into several boxes to prevent crushing. It is best to have just one layer of

cocoons to minimize the spread of mites from one cocoon to another.

The best emerging box is made of wood. The wood absorbs the sun's heat, warms up the cocoons and speeds the bees' emergence. At the same time, a wooden box protects bees from rain and predators.

4. Managing Mason Bee Emergence

Mason bees require the right weather conditions to mate and reproduce successfully. Cold, rainy spring weather can be deadly to newly emerged bees because it prevents them from flying, mating and foraging for food. (Once matured, bees can tolerate a few rainy days without feeding). They also require adequate food supply nearby. So if you have late blooming blueberry, you should delay emergence so the bees "wake up" to nearby blooms with ample nectar and pollen.

You can manage the timing of mason bee emergence by closely controlling their environment. If you keep the bees warm and moist, by using an emergence box, they will emerge. Alternately, cooling the bees can delay emergence. You will have to make an emergence box to keep the bees warm, then follow the instructions for warming and releasing the bees when the weather and food supply are right.

It is recommended to divide bee cocoons into four groups and emerge one group of bees every four to seven days. This way you won't risk the entire population should conditions turn bad. Emerged bees can be kept in the fridge for a few days, until you are ready to release them.

Create an emergence box that provides the necessary warm, moist environment. First, place toilet tissue in a plastic transparent/ translucent container. Gently lay the cocoons on top of the tissue. Cut out part of the lid for ventilation and glue screen material over hole. Make sure the screen is secure and the lid is closed tightly. Bees can escape through the smallest of holes. Place this early emergence box in a warm room held at a temperature of 22-25°C (68-77°F).

Any escaped bees should be captured and released outside. Get ready for any escaped bees by making a simple container to hold bees. Cut off the top of a plastic pop bottle and inverting it into the base. The spout should be above the base when inverted into the base of the bottle. Pick bees up with soft tweezers and place into the converted pop bottle. Soft tweezers can be made from the long metal fasteners in a Duo tang folder. When several bees have been caught release them outside.

Alternately place a warm water bottle into a picnic cooler, then place the early emergence box on the water bottle. Place damp towel over the screened lid to keep the air moist and close the cooler. Most bees emerge in the first two to three days.

In 2007 the use of an "incubation" box was described for improved emergence of *Osmia lignaria* populations (Pitts Singer et. al. 2008). Increased temperatures inside the incubation box increased the speed of emergence with a high level of successful emergence. Incubation boxes contain heating units which are set to a maximum temperature of 22°C (72°F) to increase daytime temperature to above 14°C (57°F). Results showed that emergence time is shortened and helps to synchronize bee emergence with bloom initiation.

Under controlled conditions, at 24°C (75°F) and 50-60% humidity, mason bees will emerge from their cocoon in less than a week and most of the bees will emerge within the first four days. Males emerge in the first couple of days, followed by the females (Dogterom et. al. 1999).

Do not use the oven or microwave to warm bees.

5. How to Set out Emerged Bees
When most of the bees have emerged from their cocoons, it's time to move them to their outdoor nest(s).

To be sure the bees stick close to home after their release, it is necessary to cool the bees to about 10°C (50°F). Otherwise they may fly off, never to be seen again! The bees should be released in the morning, before the temperature reaches 15°C (59°F).

To cool the bees before you release them, set the box with the remaining cocoons and emerged bees into the fridge for about 15-30 minutes. Then, place that box into another, open cardboard box. Carry everything outside and place close to the intended nest(s). Remove the lid and lay the emergence box on its side to ensure bees can walk out of box- avoiding the slippery walls. By evening, most emerged bees will have left. Replace lid snugly and put the emergence box back into the warm environment to emerge the remaining bees. Continue until all the emerged bees are released.

YOUR NOTES:

_____

_____

_____

_____

_____

_____

_____

_____

_____

_____

_____

Chapter 8
Pests, Parasites and Predators

Pests, parasites and predators are a major cause of mason bee decline where mason bees are unmanaged. Some pests attack larvae, others focus their attention on pupae or the adult stage, and some pests use every part of the bee life cycle to reproduce their own. As a result, some pests may cause a relatively minor decline in a mason bee population and others may cause a major effect. Nevertheless, if the numbers of mason bees are low, then any pest can have a major effect.

All organisms discussed below have been found in mason bee nests (Bosch and Kemp 2001). This is not an exclusive list of the organisms that prey on mason bees, but they are the more common. Mites and parasitic wasps are discussed first because these are the major pests of mason bees and other solitary bees. Although this book is written with a North American focus, different species of these pests have similar effects. For additional descriptions and coloured photographs of these North American species, see Bosch and Kemp 2001.

Other species of bees, such as the honey bee, have their own species of parasites. For example, the honey bee has two parasites, *Varroa jacobsoni* and *Acarapis woodi* mites that do not parasitize mason bees.

8.1 POLLEN FEEDING MITES
Hairy fingered mites - *Chaetodactylus krombeini*
Pollen feeding mites are a major pest of mason bees, especially in humid coastal regions. These hairy mites use every part of the mason bee's life cycle for their own food and reproduction. They are called generalists because they will invade any nest that has pollen present.

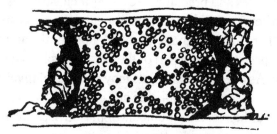

Mason bee cocoons covered in mites.

Krunic et. al. (2005) described these mites on *Osmia cornuta*. He notes the mite's reproductive cycle has five stages: egg, larva, protonymph, (hypopus), tritonymph, adult. These reproductive stages are present in the nest in spring when there is an abundance of pollen, nectar and moisture for mite development. When pollen and nectar and moisture are abundant, as many as 10 generations can be produced in one season.

A dormant stage (hypopus) appears after the protonymph stage and before the tritonymph stage. The hypopus appears in either the mobile or immobile forms and occurs when nest conditions are unfavourable, such as the end of summer and throughout the winter.

The mobile hypopus has hooks on its long legs so it can attach itself to the host. Male and female mason bees passing through mite infested cells in spring at emergence time pick up mobile hypopus and spread the mites to new nests.

The immobile hypopus has short legs. It stays in the old nest and waits for the arrival of a bee or other insects.

It is thought that mites sometimes consume bee eggs and even later stages of bee development, in addition to pollen and nectar.

1. Spring dispersal

When spring arrives mites are inside the mason bee nesting tunnel, either on cocoons or clustered between mud partitions. As temperatures rise in the spring, mason bees begin to chew their way out of their cocoons. Mites close to the chewed hole enter the cocoon and move onto the mason bee while still inside its cocoon. When the bee finally emerges, mites are visible on the bee. During mason bee mating, mites either stay on the female bee to eventually arrive at the nest, or move onto the male mason bee during the mating process. Mites that do not move out of the nest by attaching themselves to a mason bee, move out of the nest tunnel and slowly move to adjacent nesting sites. In heavily infested nests, mites can be so numerous that the bee exiting through mite infested cells, looks reddish in colour. Often the weight of mites is so great that the bee is unable to fly. Once in a new nest, the mites feed on the pollen and the cycle begins again.

2. Reproduction

The pollen lump in a nest cell is food for the new developing bee larva. In two to four days, after the mason bee egg is laid, it becomes a larva that feeds on the pollen lump. In the meantime, mites that have arrived in the mason bee compartment begin feeding on the pollen lump. Mites can deplete the pollen source and often, the larvae starves due to lack of food, or becomes food for these mites. The end result is the same. While mites feed on pollen mites begin reproducing. The production of mites can be so fast that only mites will be produced in that cell. Occasionally a small bee has been able to feed, develop and grow amongst multitudes of mites in the cell. Mites in this stage stay inside the partitioned nest and keep reproducing until the pollen food runs out or when cool winter temperatures stop the cycle.

In the fall and winter, mites are inside compartments between mud walls. Or, if the bee was able to survive and spin its cocoon, mites will also be on the cocoon.

Looking at it from a management perspective, the fall and winter months are ideal for removing mites from cocoons using the washing and screening method (Chapter 6.3). It is important that most mites are removed because any mites remaining will start the reproductive cycle again. When washing and screening is done every year, the mite level will be contained and controlled to relatively low levels. A low level is when 5-10% of nest compartments consist of mites. Even if all the mites are removed, these pollen feeding mites are part of nature and will never be completely absent inside mason bee nests.

3. Spring emergence
In a natural situation where nests are not cleaned and managed, mites become active as spring temperatures rise. Mites are ready to adhere to any passing bee emerging through the cells and mud walls. Bees passing through a compartment of mites are unlikely to survive mating because flight is often impossible due to the added weight of mites adhering to their bodies.

When a full nesting tray is examined, mite-filled compartments are present in most tunnels. For this reason, mason bees that chew their way through the tunnel to exit, have a slim chance of avoiding a mite infested tunnel. As the number of mites increases, the number of mason bees with mites increases, and the number of mites carried by mason bees also increases. Gardeners who live in humid coastal areas must manage and clean mason bee nests, or expect increasing mite infestation.

Without any management, keeping mason bees is often described as follows: "I started with bees four years ago, and the number of bees from year one to year two increased dramatically. From year two to three, there were still lots of bees, but no increase

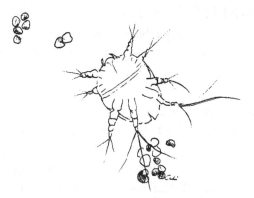

Pollen feeding mites.

was noted. Then in year four, there was absolutely no mason bee activity". The reason for this dramatic increase and crash is the typical collapse of unmanaged populations. As a wild population increases, so does the population of pests and or predators. In the case of mason bees this occurs in three to four years.

In any agricultural pursuit, the management of a crop includes keeping pest levels low, so there is a crop to harvest. This is exactly the same with mason bees. To keep mason bees, the bees have to be managed so that pest levels are at a level sustainable for bee production.

Under certain environmental and/or local conditions, keeping mason bees without management can be successful and numbers of mason bees do not appear to decline. In these situations, pests may not have optimum conditions to live and reproduce, or the availability of nests is without limit-such as buildings covered with cedar shakes. Unlimited nests would mean bees can always choose new nests.

8.2 PARASITIC WASPS

There are species of wasps that parasitize insects. These wasp species are generalists and will go after any insect host they can find and parasitize.

The parasitic wasps frequently associated with mason bees are two species of Chalcid wasps (*Monodontomerus* species and the *Melittobia chalybii*). They are both small and black. One difference is easily discernable when a mason bee cocoon is opened. The wasp of one species lays about 10 offspring (*Monodontomerus* species) and the other wasp species lays about 60 offspring.

Parasitic wasps usually overwinter in their inactive pupa stage. As the weather warms in spring, the wasp continues its development from pupa into an adult. As mason bees emerge, these wasps also emerge and begin looking for hosts such as developing mason bees. Wasps lay their eggs inside the developing bee in its cocoon stage or pupa stage. These eggs soon develop into wasp larvae. The larvae feed on the developing mason bee and kills it. Wasp larvae grow and develop into pupae and then into adults. In warm summer temperatures, wasp reproductive cycles can be repeated several times over. The number of wasp offspring in one year can be large when one considers that wasps can repeat their life cycle several times in one year and the number of offspring for each female is 10 or 60 depending on the species. So the initial hatch of 60 wasps quickly becomes 3600 wasps in one generation.

Protection from these wasps is important in late spring and summer when mason bees have stopped flying and wasps are searching for developing mason bees. Enclosing mason bee nests within wasp-proof netting can be a very effective way of minimizing the impact of these parasitic wasps (Chapter 6.4).

8.3 FUNGAL PARASITES

Fungal parasites destroy larvae or pupae of the mason bee leaving a black slightly hairy shell. This shiny black shell of the pupa can range in size from 5 mm -15 mm (3/16 - 5/8"). These fungal spore remains easily spread fungal infection throughout mason bee nests. If fungal remains are found in nests at harvesting time, do not break this shiny black shell because it will disperse fungal spores and contaminate other nesting tunnels. Set these trays aside for additional cleaning so the other nests are not contaminated. Thoroughly wash nests with bleach to remove fungal spores (Chapter 6.2.3)

If fungal parasitic remains are found in wooden nests, there are two ways of dealing with the fungi. Similar to other nests, they require a thorough bleaching. However, some fungal spores may remain inside the wooden material of these nests. Destroying wood nests that have fungal parasites is the best way of minimizing further contamination.

8.4 BEETLES

There are a variety of small and large beetles that consume mason bees in their larval, pupa or adult form (Bosch and Kemp, 2001). Beetles are generally associated with insects and insect nests where there is food, nest debris and insects to consume. Beetles are not usually a serious pest for mason bees although they can seriously decrease the number of mason bees surviving through the winter. When the number of mason bees is low, a few beetles traversing through a nest can be devastating to a bee population. Four species of beetle are commonly found in mason bee nests.

1. Checkered Flour Beetle

Just as its name implies this adult beetle is black and yellow in a checkered design. Female beetles lay their eggs at the entrance to a mason bee nesting tunnel. When opening nests in the fall it is likely you will see the feeding larval stage of this beetle. The larval stage of this beetle is bright red and it is as large as 13 mm (1/2") long. The red colour of the larva makes it easy to see amongst

Checkered Flour Beetle

nest debris. The larvae feed on: eggs, larvae and pollen provisions. They destroy the nest as they move from one cell to the next inside a nesting tunnel. After the larva feeds, it builds a chamber partition of thin brown material.

Removing nests from the garden and placing nests under cover after the pollination season minimizes contact between nests and beetles looking for nests to reproduce. Removing nests from the garden will result in fewer mason bee offspring lost to beetle predation.

2. Spider Beetle

Spider Beetles are smaller 5 mm (3/16") than the checkered flower beetles and are recognisable by their four distinct patches of white on a background of the brown wing covers. Their antennae are long. Spider beetles lay four to five eggs inside the nesting tunnel close to a developing mason bee. The larvae eat nest provisions (bee feces, pollen bee corpses and cocoons) often leaving the bee larva to starve.

Even though the adult may not be visible when examining nest contents in the fall, the mass of long fecal strings are a good indicator that larval spider beetles are present. Pupation, and development into an adult is completed in a small chamber over the winter. In spring adult beetles emerge from the nest and repeat their cycle.

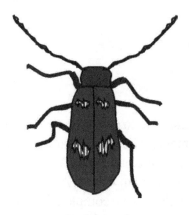

Spider Beetle

Removing these beetles from the nest when nests are cleaned is a good opportunity to keep the spider beetle population low so it does not become a serious pest.

3. Carpet Beetle
Carpet beetle adults are oval and similar in shape to lady bird beetles. Carpet beetle adults are red with blackish patches. The larvae are reddish brown, 5 mm (3/16") long and easily recognized by their long bristles on each larval segment. They feed on pollen provisions and nest debris. Carpet beetles overwinter in all stages and can have several cycles every year.

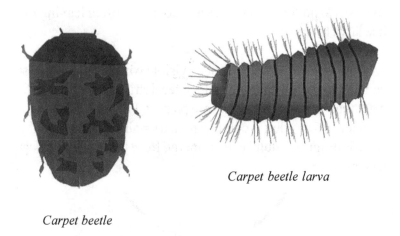

Carpet beetle larva

Carpet beetle

4. Flour Beetle
Flour beetles are similar to carpet beetles in that they also scavenge on nest contents. The adult is dark brown with grooves in the wing cover. It is a medium sized beetle (3-6 mm, 1/8-7/32"). Eggs are laid on pollen provisions. As in the other beetles' life cycles, several cycles can be completed each year. The larva of this beetle is orange/red with a dark line on each segment.

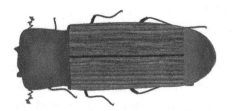

Adult flour beetle

8.5 CUCKOO BEES

Cuckoo bees (*Stelis montana*) are black, a little smaller than mason bees, and do not have white facial hairs. They enter a mason bee nest while the female mason bee is away. The cuckoo bee larva consumes the mason bee larva, pupates and overwinters inside its own cocoon. One generation is produced per year and one cuckoo bee develops in each cocoon. The cocoon can be recognized by its well defined nipple and long curly fecal pellets on the outside of the cocoon (Bosch and Kemp 2001).

8.6 SPIDERS

Spiders can consume large numbers of mason bees. Both jumping spiders and web spiders have been observed catching mason bees. Even if spiders are not visible around the nest, their webs are a good indication that spiders are feeding on your bees. Remove webs from nests to keep spider activity at a minimum.

8.7 EARWIGS

Earwigs feed on nest contents including fecal material, pollen and developing mason bees. They are particularly abundant if nests are located near wood chips and/or when wood nests are wet during spring or summer.

8.8 BIRDS

Small and large birds predate on adult and developing mason bees. Large birds such as wood peckers destroy nests by pecking them apart to get at mason bee larvae. Some of the larger woodpeckers can reach the end of a 15 cm (6") nesting tunnel with their long tongues. Other birds such as chickadees and the much larger jays have been observed picking bees out of the air while flying. Bees are particularly susceptible to getting caught by bird predators on cool days when mason bee flight is slow.

8.9 ANTS

Ants will forage on newly deposited pollen and are sometimes present when the nest is set near the ground.

Ants are a particular problem in the southern interior of the United States and more recently in California. Both the fire ant and the Argentinian ant will devastate colonies of bees, including ground nesters, above ground nesters and honey bee colonies.

8.10 MAMMALS

Both small and large mammals are known to predate on mason bee nests.

Squirrels will destroy a nest while trying to get at the bee larvae. They will chew straws or throw the nest to the ground destroying everything inside.

Bears are also known to predate on mason bee nests even though mason bee nest contents are meager pickings for a bear. There are only small quantities of pollen and nectar, but they smell the pollen inside the nesting tunnel and will chew the nest to get at this source of protein if they are hungry. Where bears are present, an electric fence is a good deterrent.

YOUR NOTES:

_____

_____

_____

_____

_____

_____

_____

_____

_____

_____

_____

Chapter 9
Mason Bees
and Their Relatives

Mason bees belong to one of seven families of bees within the Order Hymenoptera. There are six families of solitary bees within the order and one of social bees. The six families of solitary bees comprise 119 genera and 3996 species. This chapter will include an introduction to one family of solitary bees (Megachilidae) and the one family of social bees (Apidae).

A section on honey bees and bumble bees is included because I am often asked about the differences between honey bees, bumble bees and mason bees.

Both mason bees and leafcutter bees are grouped within the Family Megachilidae. Leafcutter bees are grouped into the Genus *Megachile,* and mason bees are grouped into the Genus *Osmia.* Honey bees and bumble bees are grouped within the Family Apidae.

9.1 SOLITARY BEES
Solitary bees live just as their name implies - alone in a nest and not in a social unit like honey bees. Solitary bees live in the ground, above ground, within vegetation or on bare ground.

Solitary bees can be divided into renters, excavators and builders. Renters use wood cavities, soil cavities, cracks in rocks, snail shells, plant galls and small holes burrowed by beetles or other animals. Excavators excavate into the ground, vertical banks, stems or wood. Builders build on rock surfaces, stones or stems.

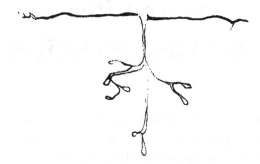

Ground nesting bees use excavated holes as their nests.

Often ground nesting bees nest in an area with suitable nesting soil characterized by its slope, texture and compactness. Thousands of bees may nest in these areas, each with its own nest. Although these bees live in close proximity, they are not social. They do not cooperate in rearing their young and thus do not share nests. Some bee species may use the same or communal tunnels, but they only tend their own nests and offspring.

1. The General Life Cycle of Solitary Bees
Adult solitary bees emerge from early spring to late summer, depending on species, elevation, climate and location.

During spring and through the summer, adults emerge from their nest and mate. Males continue mating for 1-2 weeks and then they die. A mated female then searches for a nest site and food (nectar and pollen). She provisions her nest with pollen and nectar stores on which she lays an egg. She then plugs up her nest cell and repeats this process daily until she dies.

Foragers live anywhere from a few weeks to a few months. The shorter life span is more common. Many predators, such as birds and spiders, kill and eat bees, which makes foraging a dangerous task.

Each egg develops into a larva which feeds on the pollen and nectar stores. The larva then develops into a pupa and then into an adult. Development from egg to pre-pupa occurs during the summer months. Depending on species, the pre-pupa continues its development to the adult bee (as in the mason bee, *Osmia lignaria*), or development stops and overwinters in the pre-pupa stage (as in the leafcutter bee, *Megachile rotundata*). The leafcutter bee pre-pupa continues its development to adult during the following spring.

2. Family Megachilidae

This is a very large family, present throughout the world. The females carry pollen packed into a brush of hairs called a scopa underneath their abdomen. These bees also carry material into the nest to build cell partitions and cell linings. Material may be pieces of leaves, petals, chewed leaf pulp, resin, mud, pebbles, or hairs shaved from plants. *Osmia* and *Megachile* are two genera within the family Megachilidae

3. Genera *Osmia*

The majority of the *Osmia* mason bees are renters because they nest in existing cavities. Most use masticated leaf pulp and some use mud to plug up their nest. This genus is a group of black or metallic green, blue or blue-green bees. The nesting habits of *Osmia* are diverse. Some excavate burrows in the soil while others use previously excavated burrows of other insects or even empty snail shells. Bees also use natural nesting holes such as the hollow stems of water reeds (*Juncus spp*). Other bee species create nests in hollow stems by removing the pith from plants such as elderberry and stems of the carrot family.

4. *Osmia lignaria*

Osmia lignaria is known as the mason bee or blue orchard bee and is a common, large and abundant *Osmia* species in North America. It emerges early in spring with the fruit tree blossoms. *Osmia lignaria* can be found from British Columbia to Quebec and the New England states and as far south as California, Oklahoma and Georgia. The eastern subspecies is known as *Osmia lignaria lignaria* and the western subspecies *Osmia lignaria propinqua*. *Osmia ribifloris biedermannii* is another common early spring pollinator. It has much longer hairs than *Osmia lignaria*. *Osmia ribifloris biedermannii* occurs naturally throughout the southwestern states of Oregon, California, Utah, Texas and in northern Mexico.

5. *Osmia texana*

A smaller species of *Osmia, Osmia texana*, that emerges during the summer months was studied in Victoria, Vancouver Island (British Columbia) by Rex Welland. He wrote the following description for inclusion in this book:

> "*O. texana* is a native solitary bee. It emerges at the time of blackberry bloom when the majority *of O. lignaria* have died. In Victoria, British Columbia, *O. texana* emerges in the beginning of May and remains active until early August. *O. texana* forages at warmer temperature than *O. lignaria*, and does not become active until daytime temperature reach 21°C (70°F).
>
> In the Pacific Northwest, it is approximately half the size of *O. lignaria*. But size seems to vary from one region to another. In California it is usually smaller, while in the Rocky Mountain regions it is usually larger. Females are a metallic

black/blue in colour. The male has similar colouration except that the hairs on its body are golden-red, especially in the sun. Like *O. lignaria*, *O. texana* readily accepts manmade nest boxes. The only difference is the size of the hole. *O. texana* prefers a 4 mm (5/32") diameter hole. The female provisions the chambers with a pollen-nectar mix, on each of which she deposits an egg. Like *O. lignaria*, *O. texana*, partitions off each chamber. Unlike *O. lignaria*, *O. texana* often includes slivers of wood in the final plug. Again, the stages of their life cycle are similar. But with one exception. *O. texana* are 'parsivoltine'. That is, some emerge the following season while some remain in their cocoons for a second year before they emerge. Thus, if the end-plug remains closed during the spring and summer, it may mean that these bees will emerge the following year, and not that they are dead.

It is known to visit trailing blackberry, red elderberry, boysenberry and blueberry flowers. *O. texana* can be found in British Columbia and Alberta, to Ohio and New York, and south to California (Rust 1974)."

6. *Osmia Cornuta*

This non-native species of *Osmia* has recently been introduced into the United States: *Osmia cornuta* was imported from Europe, but has not been observed. *Osmia cornuta* can be recognized by its dense black hairs on the head and abdomen and red hairs on the abdomen.

7. *Osmia Cornifrons*

Osmia cornifrons was imported from Japan and was successfully established and has become wild in parts of the United States.

8. Genus *Megachile*

The Megachile bees are nonmetallic, black bees that cut leaves and sometimes petals (e.g. rose) to line and cap each cell. There are about 115 species of leafcutter bee in North America. About

22 species occur in Western Canada.

Megachile rotundata is a species of leafcutter bee that originates from Eurasia. It is used in the commercial seed production of alfalfa. The female *Megachile rotundata* has silvery gray hairs on the underside of its abdomen. Other female leafcutter bee species usually have golden, tan, or black hairs on the underside of their abdomen.

Megachile rotundata overwinters in the pre-pupa stage. During late spring and early summer, temperatures are sufficiently warm for pre-pupae to develop into adults. Under controlled incubation conditions, development time to the adult stage is 20 to 31 days at 30°C (86°F). Adults emerge and start pollinating at temperatures of 18°C (64°F).

Commercially available nesting materials for the leafcutter bees were first made out of grooved wooden pine boards that were stacked to make a series of nests. Polystyrene is a recent innovation. Polystyrene grooved boards and nesting holes molded into solid polystyrene blocks are both used today.

9. Summer Mason Bees

There are many species of mason bees from different families and genera that pollinate for about a month, from late spring to fall. These summer mason bees nest after the early spring mason bee, *Osmia lignaria*. Nests with nesting holes that are small in diameter 0.3 mm – 0.6 mm (0.12" – 0.24") can provide homes for these summer mason bees. After several years of setting up nests, it is not uncommon to have five or six different species nesting at different times of the year.

Because there are many species of solitary bees, they are difficult to identify and few have common names. Nevertheless, using a simple description of the adult together with a description of the nest plug, the genus of the bee can often be determined, (even though the type of nesting plug is highly variable within-genus). Nest plugs are made from mud, masticated leaves, resins or a mixture of these materials (Table 1). The bee is described as either striped or metallic blue/green. In the future, the diameter of the nesting tunnel may be an additional feature used to identify a bee as to genus.

NEST PARTITION CONSTRUCTION	ADULT	FAMILY	GENERA
Polyester	Variable	COLLETIDAE	Hylaeus
Leaf cuttings	Striped	MEGACHILIDAE	Megachile
Mastics of leaves mixed with resins &	Striped	MEGACHILIDAE	Megachile
Resins, sand & pebbles	Striped	MEGACHILIDAE	Chelostomoides
Soil & or mastic of leaves and sometimes resins	Metallic (not all)	MEGACHILIDAE	Osmia
Resins and stones	Not metallic	MEGACHILIDAE	Anthidium
Pithy partitions and cap	Variable	APIDAE	Ceratina
Pebbles & Mud	Black to metallic	MEGACHILIDAE	Hoplitis

Description of some solitary bee nest partition construction and of adult bees that provides preliminary identification to genera. (after Hallett, 2001 and modified by W. Stephens; J. Cane)

In a study of gardens and natural areas, Tommasi et. al. (2001) identified five species of Osmia and seven species of Megachile in Vancouver, B.C., Canada. All these species could easily coexist in one garden.

9.2 SOCIAL BEES

Social bees such as honey bees and bumble bees are different from solitary bees because the insects work together to reproduce. They are present in large numbers in a nest, they gather food, defend the nest and care for their young. Examples of social insects are wasps and all ants and termites.

1. Honey bees

Honey bees live together in a colony that may reach a size of 60,000 bees in the summer. There is one queen that lays the eggs of a colony. The remaining bees are workers and drones. Workers are underdeveloped females that are not fed the large amounts of royal jelly given to developing queens. Drones are males.

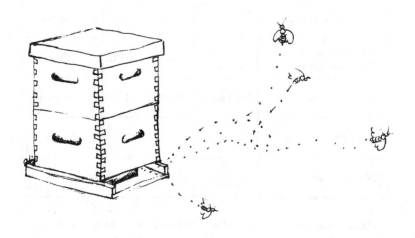

Honey bees and their hive.

Honey bee workers do all types of tasks. Inside the hive, young bees begin by tending and feeding the young brood and cleaning out cells. As workers age, foraging becomes their major task. Depending on the needs of a colony the workers become foragers for pollen, nectar, water, or plant resins (propolis). They continue their foraging tasks until they die. In the summer it can be as short as 14 days because foraging is a dangerous task. Birds, insects, spiders, skunks, bears and humans all prey on bees and their nest contents. Unlike the workers, the males of a colony, or drones, spend most of their lives inside the colony. On maturity, they fly out of the nest to find virgin queens. New queens mate on the wing with several drones. Drones die after mating.

2. Bumble bees

Bumble bees also are social bees. A colony of bumble bees works together as a cohesive unit to find food, store food and produce workers, queens and drones. Like honey bees there is usually one queen, but bumble bee hives are smaller with only a few hundred workers.

Bumble bees are important early spring pollinators. Because of their size and 'fuzziness', they do an excellent job of transferring pollen from one flower to another. Different bumble bee species appear at different times throughout spring, but because colony life extends over several months, several species may be present at the same time of the year.

In nature, bumble bees may nest underground, in a cavity alongside roots, or in a tree cavity created by other animals. In suburbia, bees nest in walls with fibreglass insulation or in cotton bedding of abandoned couches and mattresses. Bird boxes with remains of old bird nests are also used by bumble bees as are abandoned mouse nests.

Bumble bees emerging from hibernating hole.

*Bumble bees nesting in a hollow
amongst tree roots.*

When blossoms first appear early in spring, a large mated bumble bee queen emerges from her hibernation cavity. She will first search for nectar to replenish her energy. Then, she searches for a summer nest. When she has found a nest site, the new queen collects nectar and pollen from flowers. At the nest, she mixes pollen with nectar and forms a lump of food for her young. When the pollen lump is adequate, she lays a cluster of eggs on this pollen lump and buries them within the surface of the pollen lump. The pollen lump may be as large as the first joint of your thumb. Then she begins a 20-30 day brooding period. Bumble bees generate heat by vibrating their thoracic muscles. Heat produced keeps the eggs warm and the eggs then develop into larvae. The larvae feed on the pollen-nectar lump. Each larva then transforms into pupa and emerges as an adult bumble bee. This transformation from egg to adult takes about 16-25 days. Poor food in the early spring can result in small offspring.

By early summer, the number of workers may reach 150 individuals. The largest known bumble bee colony of the temperate zone (east coast of North America) had approximately 3,000 individuals. In the middle of the summer, when the colony is large, the colony produces new queens and drones. About one month later, virgin queens emerge from the nest, feed on nectar and mate with the drones. Once drones emerge they do not re-enter the nest. They stay over night on and among flowers and foliage. Drones can be recognized by abundant yellow hairs on their faces, longer antennae and lack of pollen baskets.

Come fall, the newly mated queens find a hibernation cavity and remain there until the following spring. The location is often on a slope to prevent the drowning death of hibernating queen bees through the wet winter months. The hibernation hole may be a very small cavity in soil, peat moss, or similar. When all queens leave the colony, the original colony eventually dwindles, and the remaining worker bumble bees die.

YOUR NOTES:

Chapter 10
Novel Ideas

Following the thread of a novel idea can be an exciting adventure. Testing the idea, modifying it and improving it, is all part of the adventure. This chapter, includes some of these ideas, stories and experiments from mason bee keepers.

10.1. Black Nests
B. Smith, Coquitlam B.C. Canada

Bob Smith noticed early on that mason bees prefer black nesting tunnels. He used grooved trays, and painted the whole tray with water based black paint. He did not know why they preferred the black trays except that black wood would be warmer than brown wood. Warmth of course helps speed up development of young mason bees.

10.2. A Backyard Management System
C. Saboe, Surrey, B.C. Canada

"This year I went all out on paper straws which I colored with 10 different colors. Paper straws have quite a few faults. First of all they will absorb the nectar which is vital for pupa development. The parasitic wasp is capable of drilling through two layers of paper. Paper straws have to be replaced every year and have to be colored at one end. Whereas the wooden housing can be cleaned and re-used.

I make the wooden housing out of 1"x4" spruce or pine and can be machined at random length and then cut into even 5 1/2" length and then stacked to any height you want. I put duct tape to cover the holes at the back. Then I wrap duct tape all around the top, bottom and two sides. No need to use screws, the tape holds them together. It makes it water proof and keeps the light out and I don't believe that wasps can drill through them (it's sticky). I paint all the holes (ends) 10 different colors. I use water color which only lasts one year. Last year I colored the wooden housing. Color does help and keeps the little bees from wondering elsewhere. I had 2500 mason bee cocoons last year. I expected 10,000 cocoons for this year but the water got in with the heavy rain and we only had 7500."

10.3. Feeding Mason Bees
S. Dupey, Twisp, W.A., U.S.

"If cool spring weather limits bee activity, probably, a feeding station near the nests would substantially increase the nesting rate of emerging bees in addition to their overall number of cells provisioned. This seemed to be the case in my experiment with feeding *Osmia lignaria* one season.

My simple feeder consisted of an upside down ma-

son or mayonnaise jar set upon a plate, pooling nectar there due to small slots cut in the plastic screw-on
lid. The sugar solution reached a level which stayed
more or less stationary due to the chicken waterer
principle where more nectar leaks out once the level
drops enough to allow air into the slots cut in the lid.
A doughnut shaped screen (preferably plastic or
stainless steel) with about 1/8" mesh, covered the
plate and nectar, and the hole cut in its center fit snugly
around the upside down jar lid. A snug fit at the screen
margins was essential to keep bees from finding their
way into and drowning in the solution.

Stubby stainless steel bolts were affixed as legs pro-
truding from the upside down lid. Adjusting the nuts
for greater or lesser protrusion (about 1/4" inch being optimal) allowed one to regulate the depth of the
nectar pool under the screen by raising or lowering
the jar slightly. This was somewhat important, as too
deep would flood the screen, while too shallow might
be out of reach of the mason bee's probing tongues.

Initially, Osmia or other solitary bees probably won't
feed very readily at such a feeder without some en-
ticement and conditioning. I enticed bees to visit the
feeder by positioning it directly in front of the nests
(within a few yards and in an ant-proof location).
The bees were conditioned to feed by scattering some
fresh flower petals upon the screen each morning and
spraying these with some of the sugar-nectar solution. Bees landing to investigate would usually discover the nectar droplets after some probing. Further feeding and probing through the screen resulted

in regular visitation and feeding from the underlying nectar pool. The visiting bees themselves seemed to attract others in growing numbers and eventually this small feeder was used by large numbers of my mason bees, usually as they were leaving the blocks to forage.

A drawback to the system was the need to renew fresh flower petals everyday or so to keep up a high level of interest and visitation from the bees. Colored and scented attractants could probably be devised.

The sugar solution would tend to crystallize more and more over time and washing the plate and jar with hot water periodically was a good practice. Also the jar would heat up in sunlight and then increased air pressure would force more sugar solution out of it and sometimes flood the screen. A foam insulating cover over the jar helped somewhat, but perhaps a better solution would be to experiment with a float-switch and an external feed solution reservoir.

Energy levels for bees using this feeder seemed significantly enhance and presumably allowed them to range farther and pollen forage more effectively."

10.4. Mite Control by Immobilization
by D. and J. Scarth, Vancouver, B.C. Canada

"A major advance in mite control was made in the 1990s when drilled wood-block nests were replaced by nesting tubes and nesting trays. With tubes and trays the cocoons can be removed from the nest in the fall and thus the bees avoid having to crawl out through mite-infested cells in the spring. Another benefit of cocoon extraction is that mites can be removed from the surface of the cocoon by washing and /or screening.

It has proven very difficult to rid the cocoons of all surface mites and until recently they were not thought to be a problem as it seemed unlikely that mites would have much opportunity to attach themselves to the bee after it had emerged from the cocoon. However, some bees that had just emerged were observed to have a cluster of mites in the 'waist' area which was puzzling because these mites could only have come from the surface of the cocoon. This was confirmed by watching several emergence sequences under a microscope, beginning with the initial breakout through the cocoon and ending some 4 to 5 minutes later with bee emergence. As the hole was being enlarged, mites were observed to crawl rapidly over to the hole and enter the cocoon. Incredibly, before the bee emerged the mites had instinctively attached themselves to the waist area where they were least likely to be rubbed or brushed off.

Clearly, some way had to be found to immobilize the mites on the cocoon surface without making it more difficult for the bee to break through. The solution was to spray the cocoons with a non-aerosol hairspray just before they were placed out with the nests. The experiment was carried out on about 300 cocoons as described below. Based on this admittedly limited experiment it was concluded that first, the spray was very effective in immobilizing the surface mites, and second, the spray as applied in this experiment presented no serious barrier to the emerging bees as there was 98% emergence.

The cocoons used in this experiment had been given a primary wash in fresh water early in the year, and a screening to remove surface mites in late April in preparation for deployment. The cocoons were sprayed in the same 8" wire sieve that was used for screening, in three lots of about 100 cocoons each. Care was taken to arrange the cocoons in a single layer in the bottom of the sieve. Spraying consisted of two or three short "whiffs" of hairspray from about 8" above and 8" below the cocoons. While spraying, the objective that was kept in mind was a thin, even layer of spray on each cocoon. Care was taken not to over-spray as this might trap bees in their cocoons."

Chapter 11
Projects

A few fun projects for kids and adults.

11.1 MAKING YOUR OWN NEST
Suitable for kids 8 years and older
Making your own nest is fun. Kids can help with rolling the nesting tubes, bundling nesting tubes and painting the ends of the nesting tubes.

This project consists of making a wooden box with 4 sides and a back, and rolling paper nesting tubes out of newspaper.

1. What you need for the nesting box:
12 galvanized box nails
1 hammer
1 screw and screwdriver
1 hanger
Sand paper
5 pieces of wood about 1 cm thick:
- Left top 124 mm x 209 mm (4.9" x 8.2")
- Right top 141 mm x 209 mm (5.6" x 8.2")
- Left bottom 95 mm x 188 mm (3.7" x 7.4")
- Right bottom 109 mm x 188 mm (4.3" x 7.4")
- Back 95 mm x 92 mm (3.7" x 36")

2. What you need for the nesting tubes:
1 newspaper, cut into strips
8 mm (5/16") dowel (wood or metal)
1 roll of masking tape
Several sheets of plain white paper or Kraft paper

3. How to make the nesting box:
- Assemble nest box by nailing sides and top with 8 nails; two per side.
- The back is secured with 4 nails.
- Screw hanger onto the top part of the back.
- Sand rough edges with sand paper.

4. How to make nesting tubes
Cut the newspaper and Kraft paper into matching strips 15 cm (6") wide by 61 cm (24") long. Before you roll the tubes, be sure to place the Kraft paper on top of the newspaper. The Kraft paper must form the inside of the tube. This way toxic ink from the newspaper won't come into contact with the developing bee.

Place the dowel on the end of the strip and roll, right to the end of the strip, and secure with masking tape. Ease the dowel out of the newly formed tube. Then fold down one end of the tube so you can't see any light when you look down the open end.

Bundle the tubes together and place them in a nesting box. Hang it on the east side of a house or shed, about 1.5 m (5') off the ground.

11.2 MASON BEE POP BOTTLE
by Rex Welland

This nest is constructed using a modified 2 litre "pop" bottle for the main structure and rolled newspaper for the nesting tubes.

This 'pop bottle' nest has many features. The tubes can be easily removed and unwrapped. This provides easy access to the cocoons for fall cleaning. When the parasitic wasps try to sneak down between the tubes and masking tape, they get stuck on the tape at the front end of the tube bundle - like 'flypaper'. The

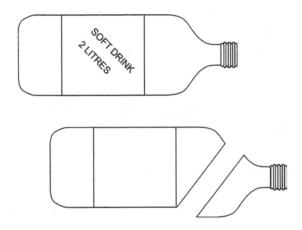

extended and sloped 'roof' of the bottle provides excellent weather protection. The extended sides provide not only a wind break for the bees but also seem to somewhat deter wasp parasites from entering. The container (with new tubes) can be used for many years.

1. Nest Container Construction:

1. Remove the product label.

2. Cut off the neck of the pop bottle at an angle as shown

3. Lightly abrade (sand) the exterior of the bottle to provide a good bonding surface for paint. The exterior needs to be painted to keep the light out.

4. Paint the bottle with primer. Ensure all exterior surfaces are well covered, including the bottom.

5. If you wish to add a plywood mounting bracket, now is the time to do it - otherwise skip to step 8. The bracket is a 110 mm x 170 mm (4 1/4 x 6 3/4") piece of 1/4" plywood.

6. Sand the primer off the 'feet' on the bottom of the bottle (where it will touch the plywood). This will ensure a better bond between the plastic of the bottle and the plywood.

7. Hot glue the bottle to the plywood.

8. Apply a finish coat of paint, in the colour of your choice, to the bottle and the plywood.

2. Tube Construction:

9. Open a newspaper right out flat (2 pages wide). Fold it from top to bottom twice. Run a knife along the folds. This will create strips of paper approximately 15 cm (6") wide and as long as the newspaper.

10. Starting at one end, tightly roll one newspaper strip one at a time around an ordinary wooden pencil to form a tube. Secure the loose edge of the paper with a couple of pieces of masking tape. Repeat the process to create as many tubes as you like.

11. Bundle about 30 tubes together. Using a couple of elastic bands to hold the tubes together, glue a piece of light cardboard on one end. Wrap a piece of wide masking tape around the tubes at the other end. [Note - do not be concerned if the tubes are slightly different lengths]

12. Put some upholstery cotton (unbleached cotton) in the bottom of the 'bottle' (remnants or scraps should be available at your local upholstery shop). Insert the tube bundle, cardboard end first. Put some more upholsters cotton in between the tubes and the sides of the bottle.

13. Mount the completed Mason Bee nest box on an east facing fence or wall in the sun.

YOUR NOTES:

_____

_____

_____

_____

_____

_____

_____

_____

_____

_____

Chapter 12
Bee Management Throughout the Year

A guide to mason bee management.

The bee management chart on the following page provides a simple guide to managing your mason bees. Each year, make notes and amend this chart to suit your region and management plans.

Month	What to do:
JANUARY	Assemble nests.
FEBRUARY	Set nests outside, refrigerate cocoons until released.
MARCH	When bloom is a week away, set cocoons outside for release.
APRIL	Add additional nests, according to blossom presence. Plant early flowering plants that bloom when there are few flowers present.
MAY	Protect from rain.
JUNE	Protect from predators with chicken wire.
JULY	Protect from parasitic wasps with netting.
AUGUST	Protect nests from extreme heat.
SEPTEMBER	Protect nests from rain.
OCTOBER	Collect nests and harvest cocoons. Clean nest and cocoons. Store cocoons.
NOVEMBER	Check cocoons for mould and predation.
DECEMBER	If extreme cold temperatures persist, move cocoons inside.

Afterword

The art and science of keeping mason bees will continue to evolve. Every year hobbyists create unique tools and methodologies for making this hobby within reach of most gardeners.

As more and more people come on board this fascinating hobby, more ingenious ideas will be developed for maintaining mason bees and fruit production.

These bees make a wonderful hobby. Best of luck with your discoveries and fruit production

If you have devised a new method that makes keeping mason bees easier or more interesting, and you would like us to consider your story in the next edition — please contact us.

Feed back for the next edition is always appreciated.

You can email us at info@beediverse.com

YOUR NOTES:

Glossary

Alfalfa leafcutter bee. An introduced Eurasion species *Megachile rotundata*, belonging to the leafcutter family Megachilidae, that forms the basis for a multi-million-dollar agribusiness for pollinating alfalfa. The bee is easily managed in straws, boards, and styrofoam nesting blocks.

Anther. The pollen containing part of the floral stamens where the pollen grains are produced.

Biodiversity. The variability of living creatures within a habitat.

Blossom. A flower that is open for pollination.

Blue Orchard Bee. Common name for *Osmia lignaria*

Bumble bee. The common name for any bee in the *Genus Bombus*. These social bees are large hairy, often black, white and yellow, or reddish.

Cocoon. The protective waterproof envelope spun by larvae and used as a covering during the winter pupal stage.

Corn-Material. Biodegradable, environmentally friendly material made from corn. A renewable resource

Cross-pollination. The transfer of pollen from the anthers of one plant to a recipient stigma of another.

Grooved channels. Channels are created by cutting into one side of a piece of wood using a router or table saw.

Hive. A structure for housing honey bees.

Honey bee. A social bee that nests in a colony and produces honey.

Hymenoptera. The order that includes bees and ants.

Incubator. A warming box used for emerging mason bees.

Larva(e). The newly hatched and wingless worm-like feeding stage of many insects.

Mason bee. The common name given to any bee in the *Genus Osmia* and belonging to the Megachilidae bee family. It is called a mason bee because they plug their nest holes with mud.

Megachilidae. Large bee family including leaf-cutting bees and mason bees.

Mites, Varroa and Tracheal. Two parasitic mites of the honey bee.

Nectar. A liquid mixture of sugars secreted by the nectaries of plants.

Nectar-Pollen lump. Also known as Pollen lump. A mixture of pollen and nectar placed in the nest to feed larvae.

Nest. A nesting tunnel which is provisioned with food for developing young.

Nesting box. A structure that protects nests from the elements.

Nesting channels. Grooves in wood created by a router or table saw.

Nesting tubes. Tubes of paper or cardboard used by mason bees for nesting.

Orchard Mason bee. Alternative name for Mason bee, *Osmia lignaria.*

Parasite. An animal that produces its young inside the host, and thereby destroying its host.

Pollen. Microscopic particles in the anthers of flowers that contain the male sperm nuclei.

Pollen tubes. Microscopic tubes produced by the pollen grains that connect them to the ovary of the flower. The genetic material of the pollen grain travels to the ovary via the pollen tubes.

Pollen-feeding mite. A mason bee pest that is particularly prevalent on the northwest coast of North America.

Pollination. The process of moving pollen from the anthers to the stigma of either the same flower or another.

Predator. An animal that feeds on another.

Pupa: The resting stage of insects after larva and before adult.

Queen. The principle or only egg-laying female in a social colony, that does little or no foraging.

Social bees. Bees that live together in a communal nest and share foraging and nest duties. Honey bees and bumble bees are social bees.

Solitary bees. A female bee that lives alone in her own nest.

Stigma. The female receptive portion of the style.

Stamen. The male structure bearing the pollen grains of a flower.

Style. The middle connective portion of a female flower.

Tray. A piece of wood or plastic with grooves. When stacked, trays create nesting tunnels for mason bees.

Yurt. A six or eight-sided field structure used for protecting mason nests from the weather. It is covered with tarp, has one side open and has a central hole in the roof for ventilation and access by bees.

References and Guide to Further Reading

This guide to further reading includes references listed in the text. Italicized *text* after the title is a brief description of the book or article. The title of each books is set in **bold**. The presence of only black and white photos [*b&w*] or only coloured photographs [*colour*] are noted for books only.

Alford, D.V. 1978. **The Life of the Bumble bee**. Davis-Poynter, London. *A naturalists' overview of bumble bees. [b&w]*

Banaszak, J. and L. Romasenko. 1998. **Megachilids Bees of Europe.** Wydawnictwo Uczelniane WSP w Bydgoszczy. [in English] ISBN: 83-7096-268-8. *Detailed descriptions and technical sketches of Megachilid solitary bees. [b&w]*

Bosch, J. and W.P. Kemp. 2001. **How to Manage the Blue Orchard Bee, Osmia lignaria, as an Orchard Pollinator.** Sustainable Agriculture Network (SAN), Washington, DC. [*available free online, http://www.sare.org*]

Bosch, J. and W.P. Kemp. 2004. Effect of pre-wintering and wintering temperature regimes on weight loss, survival, and emergence time in the mason bee *Osmia cornuta* (Hymenoptera: Megachilidae). Apidologie 35:469-479.

Buchmann, S.L. and G. P. Nabhan. 1996. **The Forgotten Pollinators**. Island Press / Shearwater Books, Washington DC. *Stories from around the globe emphasizing the importance of pollinators.*

Dogterom, M.H. 1999. **Pollination by Four Species of Bees on High Bush Blueberry.** PhD thesis. Simon Fraser University.

Fabre, J.H. 1916. **The Mason- Bees**. Dodd, Mead & Co. NY.

Forey, P. and C. Fitzsimons. 1987. **An Instant Guide to Insects**. Gramercy Books, NY. *How to identify insects. [colour]*

Free, J.B. 1993. **Insect Pollination of Crops**. Academic Press, London. *Technical details on the pollinator requirements of crops. [b&w]*

Griffen, B. 1993. **The Orchard Mason Bee**. Knox Cellars Publishing, Bellingham.WA. *An introduction to keeping orchard mason bees. [b&w]*

Guetdot, C., J.Bosch and W.P. Kemp. 2007. Effect of three dimension and color contrast patterns on nest localization performance of two solitary bees (Hymenoptera: Megachilidae) Journal of Kansas Entomological Society 80: (2) 90-104.

Guedot, C., T.L. Pitts Singer, J.Buckner, J.S.Bosch and W.P.Kemp. 2006. Olfactory cues and nest recognition in the solitary bee *Osmia lignaria*. Physiological Entomology 31:110-119.

Hallet, P.E. 2001. A Method for 'Hiving' Solitary Bees and Wasps. American Bee Journal 141(2): 133-136.

Heinrich, B. 1979. **Bumble Bee Economics**. Harvard Univ. Press, Cambridge, Mass. *Overview of bumble bees and their lives. [b&w]*

Henderson, C.L. 1992. **Landscaping for Wildlife.** Minnesota's Bookstore St. Paul M.N. *Detailed information on how to make our land more attractive to wildlife. Extensive list of good food plants for bees, moths, birds and other animals. [colour]*

Kemp, W.P. and J.Bosch. 2005. Effect of temperature on *Osmia lignaria* (Hymenoptera: Megachilidae) prepupa-adult development, survival, and emergence. Journal of Economic Entomology. 98(6):1917-1923.

Krunic, M., L. Stanisavljevic, M. Pinzauti, A. Felicioli. 2005. The accompanying fauna of *Osmia cornuta* and *Osma rufa* and effective measures of protection. Bulletin of Insectology 58(2): 141-152.

Mitchell, T.B. 1962. **Bees of the Eastern United States Volume II**. The North Carolina Agricultural Experiment Station. *Technical book on how to identify families Megachilidae, Anthophoridae, Xylocopidae and Apidae. [b&w]*

McGregor, S.E. 1976. **Insect Pollination of Cultivated Crop Plants.** Agriculture handbook. **No. 496**. United States Department of Agriculture, Washington, D.C.

Michener, C.D., R.J. McGinley and B.N. Danforth. 1994. **The Bee Genera of North and Central America.** Smithsonian Institution Press, Washington. *Technical book on how to identify bees according to genera. [b&w]*

Michener, C.D. 1974. **The Social Behavior of the Bees** – A Comparative Study. Harvard University Press, Cambridge, Mass. *A technical book that compares the behaviour of social and solitary bees. [b&w]*

Muller, A., A.Krebs and F.Amiet. 1997. **Bienen.** Weltbild Verlag GmbH, Augsburg. ISBN 3-89440-241-5 [in German]. *Clear & vivid photographs of nesting solitary bees, their nest structures and habitat. [colour]*

Norden, B.B. 1991. **The Bee.** Stewart, Tabori & Chang Inc. NY. *A pop-up book on social and solitary bees for adults and children. [colour]*

O'Toole, C. and A.Raw. 1991. **Bees of the World.** Blandford Publishing, London. UK. *Excellent overview of bees, both solitary and social. [colour]*

Pitts Singer, T. P., J.Bosch, W.Kemp and G.Trostle. 2008. Field use of an incubation box for improved emergence timing of *Osmia lignaria* populations used for orchard pollination. Apidologie 39: 235-246.

Richards, K.W. 1984. **Alfalfa Leafcutter Bee Management in Western Canada.** Agriculture Canada Publication No. 1495 / Economics Communication Branch, Agriculture Canada, Ottawa. *Overview of commercial management of alfalfa leafcutter bees. [b&w]*

Rust, R.W. 1974. **The Systematics and Biology of the genus *Osmia*, Subgenera *Osmia, Chalcosmia*, and *Cephalosmia* (Hymenoptera: Megachilidae).** The Wasmann Journal of Biology. 32:1-93.

Seeley, T.D. 1995. **The Wisdom of the Hive:** The Social Physiology of Honey Bees. Harvard University Press, Cambridge, Mass. *In-depth look at the behaviour of honey bees. [b&w]*

Stephen, W.P. 1957. **Bumble Bees of Western America (Hymenoptera: Apoidea).** Technical Bulletin 40. Agricultural Experimental Station, Oregon State College. Corvallis. *Technical book with range maps and line drawings of colour variations of some species.*

Tommasi, D., A. Miro and M.L.Winston. 2001. **Bee Diversity and Abundance in an Urban Setting.** Simon Fraser University, Burnaby, B. C. Canada.

Von Hagen, E. 1994. **Hummeln.** Weltbild Verlag GmbH, Augsburg. [in German] *Excellent colour photographs of habitat, nests and bumble bee species. [colour]*

Wei S., R.Wang, M.J.Smirle, H.Xu. 2002. Release of *Osmia excavata* and *Osmia jacoti* (Hymenoptera: Megachilidae) for apple pollination. Canadian Entomologist 134:369-380.

Winston, M.L. 1987. **The Biology of the Honey Bee**. Harvard University Press, Cambridge, Mass. *An overview of honey bee biology. [b&w]*

Index

YOUR NOTES:

_____

_____

_____

_____

_____

_____

_____

_____

_____

_____

_____

_____

YOUR NOTES:

_____

_____

_____

_____

_____

_____

_____

_____

_____

_____

_____

_____

_____